ECOLOGY OF BEHAVIOUR

RECENT ADVANCES IN ENVIRONMENTAL ECOLOGY SERIES—9

Ecology of Behaviour

A. P. DIWAN
D. K. ARORA

ANMOL PUBLICATIONS PVT LTD
NEW DELHI-110 002

ANMOL PUBLICATIONS PVT LTD
4374/4B, Ansari Road
Darya Ganj, New Delhi-110 002

First Edition, 1995

ISBN 81-7488-118-2

Printed in India

Published by J. L. Kumar for Anmol Publications Pvt. Ltd., New Delhi and Printed at Amit Enterprises, 518/14, Anup Mkt., Maujpur, Delhi.

Preface

In considering the behaviour of animals in the natural environment, it is important to have some understanding of the effects of the consequences of the behaviour on the survival of the animal. The consequences of a particular activity depend largely upon the animal immediate environment. If the animal is in a situation to which it is adapted, the consequences of the activity may be beneficial. The same activity performed in a different environment may be harmful. In order to study the behaviour of animals one needs to understand the adaptive relationship between the animal and its environment.

Often it is fairly easy to establish that a behaviour pattern is useful to an animal—helping it to acquire food or water, avoid an enemy, secure a home, find a mate, or protect its offspring. It is true that there are puzzling behaviours, but even these can usually be correlated with an ecological problem if the observer is patient and inventive.

The material for this book has been drawn from various information sources and it deals with the different ecological and behavioural patterns that are normally observed in man and animal.

The authors have freely consulted a large number of foreign books and Journals in the preparation of the present title and also to make it comprehensive and uptodate and to that extent it is not claimed to be our original work.

Any worthwhile criticism and suggestion for improvement would be thankfully acknowledged.

AUTHORS

Contents

1

Ecology of Behaviour

In considering the behaviour of animals in the natural environment, it is important to have some understanding of the effects of the consequences of the behaviour on the survival on the animal. The consequences of a particular activity depend largely upon the animal immediate environment. It the animal is in a situation to which it is will adapted, the consequences of the activity may be beneficial. The same activity performed in a different environment may be harmful. In order to appreciate have the behaviour of animals has been shaped during evolution, we need to understand the adaptive relationship between the animal and its environment.

Often it is fairly easy to establish that a behaviour pattern is useful to an animal—helping it acquire food or water, avoid an enemy, secure a home, find a mate, or protect its offspring. It is true that there are puzzling behaviours, but even these can usually be correlated with an ecological problem if the observer is patient and inventive. Mary Seely and W.J. Hamilton observed trench digging only at night just before or during one of the heavy fogs that roll across the sand dunes at times in this area. The trenches were typically built at right angles to the wind so that moisture-laden air collided directly with the ridge of sand. Measurements showed that tne moisture content of the ridge sand was higher than that in surrounding flat sand dune. The beetles patrolled their trenches touching the sand with their mouthparts; after a beetle had walked along its trench, the moisture content of the sand in the ridge was reduced. The inescapable conclusion is that these beetles build trenches to trap moisture from fogs, which they drink.

An evolutionary approach requires that we ask why a behavioural trait might elevate inclusive fitness more than any alternatives. Other desert-dwelling insects drink at seeps and springs or they chew detritus that has absorbed moisture from the air during a foggy or humid period. There are no sources of groundwater in vast areas of the Namib Desert, and this suggests why the tenebrionids do not search for pools of surface water. However, moisture-soaked detritus is available after a fog and is utilized by other insects. Why do the beetles dig trenches instead of eating wet vegetable debris?

PHYLOGENETIC INERTIA

In the absence of variation, selection cannot occur; if an alternative never originates, it will not have a chance to be selected for and spread through a population. Most evolutionary biologists feel that arguments based on Phylogenetic Inertia are arguments of last resort. The diversity within any category of animal behaviour is in most cases high, even among related species. Therefore, it seems fairer to assume, at least as a working hypothesis, that many alternatives have arisen by mutation within any species and that the option practiced by members of a current population has out competed the alternatives. A classic test of the ecological significance of a behaviour pattern was Niko Tinbergen's experiment on egg-shell removal by blackheaded gulls. Shortly after its chick was hatched, a parent guarding the nest takes the eggshell in its beak and flies off with it to drop the shell some distance away. Tinbergen suggested that the white interior of the shell might attract predators to a nest where they would find and consume the recently hatched young or any unhatched eggs. He tested this idea experimentally by placing unhatched eggs in a meadow visited by foraging carrion crows. By some eggs he placed a broken eggshell; by others he did not. The crows found and consumed a much higher proportion of the eggs that had eggshells nearby than of those that did not. The experiment supports the idea that at least one function of disposing of broken eggshells is to lower the risk that the parent's eggs or young will be discovered by a visually hunting predator.

It is difficult to carry out experiments of this sort. Moreover, each test can contrast only a limited number of all possible alternatives. Tinbergen contrasted the possibilities of eggshell removal versus no eggshell removal. Actually, the gulls wait an hour or so before taking the empty shell away. This may give the young chick time to dry its feathers and become partly mobile and thus better able to escape if attacked while its parent is disposing of the eggshell. Thus a more complete comparison might be between eggshell removal (1) immediately after hatching versus (2) 1 hour after hatching versus (3) 4 hours after hatching. This wouid be a difficult experiment to conduct, and because the difference in success of the alternatives would almost surely be slight, one would have to invest enormous amounts of time and energy to collect a sufficiently large sample if one hoped to find statistically significantly large sample if one hoped to find statistically significant differences among the three conditions. Fortunately, one need not rely exclusively on experimental tects to make firm statements about the adaptive value of behavioural traits. Making the right kinds of comparisons among species can provide a surprising amount of information about the function of behaviour.

THE USES OF DIVERGENT AND CONVERGENT EVOLUTION

There are two kinds of comparisons that are most productive in studies of behavioural ecology: a search for behavioural differences among closely related species and a search for similarities among unrelated species. Close relatives usually exhibit similar behavioural traits because they share common ancestry and, therefore, have presumably inherited similar genetic-developmental mechanisms. However, in some instances there are striking differences in the behaviour of closely related animals. If it is possible to identify a clear-cut difference in the environmental pressures acting on the species, one can then make a strong case that the behavioural differences between them are attributable to Divergent Evolution. Sharply different selection pressures acting on related species can cause them to diverge.

On the other hand, unrelated species could be expected to behave differently because they have different ancestors and distinct genetic-developmental characteristics. Yet sometimes two species with different evolutionary histories are behaviourally similar. These unexpected similarities may be attributable to Convergent Evolution—the evolution of analogous traits in unrelated species as a result of shared selection pressures. Careful examination of the environments of these species may reveal a common ecological pressure that can be related to the trait in question. If it does, this strengthens the argument that the adaptation has evolved in response to this environmental factor.

THE ECOLOGY OF GULL

The behaviour of black-headed gulls is typical of the vast majority of gulls. This species nests in colonies on the ground in grassy sand dunes. In the spring a black-headed gull often, but not always, returns from its wintering grounds to the nesting colony in which it was born (if this is its first breeding year); if it has already nested in a particular site, it is almost certain to return there to breed again. Males arrive before females. They take up a position in the dunes and repeatedly utter a characteristic loud raucous call (the long call), given as the bird stretches forward obliquely. It is unusual for a male giving the long call to be aproached by another male, but if this should happen, the caller may adopt a position called the upright display, in which the gull stretches its neck up, holding its beak ready to stab downward. If the other bird persists, the gull may then perform a forward display with its body held parallel to the ground and its beak pointed directly at the opponent. This display is often followed by a charge in which the bird attempts to strike the other male with the beak and wrist of the wing. Almost always the intruder is driven from the area. Examination of a colony day after day reveals that the same male returns repeatedly to the same area and will defend his territory against all comers.

FIG. 1.1: Black-Headed Gulls.

A female circling over the colony is attracted by long-calling males. She comes closer and alights in the place a male defends from other males. Initially the male may drive the female off as well, and she appears quick to leave the spot. However, a female may return to a male over and again. Eventually, the two birds may begin to engage in a number of mutual displays, such as the parallel-forward display (both birds walk side by side with their bodies stretched forward) and the facing-away display. These occur very frequently early in courtship, but gradually fade away and are replaced by food begging by the female. In response, the male regurgitates items he has collected, and his mate feeds upon them. Finally, copulation is preceded by a mutual head-tossing display, in which both birds walk about together jerking their heads up repeatedly while uttering soft calls.

If courtship was conducted in a preterritory, an area outside the nesting part of the colony, the pair searches for a place to build a nest in the breeding area. They settle in a spot, defend it vigorously from their neighbours with a great deal of raucous bickering, build a nest, and raise a brood. The nest is a simple loose mound of grasses and straw collected from the area. A pocket is made when the bird settles into the mound. After the female completes her clutch, incubation is shared by both birds. The eggs of gull are well camouflaged against the grass, saud and twing around the nest.

When the young hatch, they beg in much the same fashion as herring gull chicks, pecking at the tip of the beak of a parent. Both adults share in the regurgitation feeding of their offspring, with one bird protecting the nest while the other is out searching for food. The adult that had the first watch after the eggs hatch will wait for an hour or so and then pick up the empty eggshells, fly off, and drop them some distance from the nest. As the chicks grow older and wander more, adults give an attraction call; the chicks respond by running back to their parents to be fed. Throughout the entire reproductive period, black-headed gull adults are extremely sensitive to the presence of predators—larger guils, foxes, badgers, crows, hawks, and humans. At the slightest disturbance birds begin to give an alarm call. Young birds freeze in the nest or, when they are older, dart for cover. Adult birds become alert and vigilant. If a predator is spotted, groups fly toward it calling noisily and defecating profusely while dive bombing the enemy. As the young birds grow older, feedings reach a peak and then decrease as the fledglings near independence. Soon they are left to fend for themselves, and in the fall young and old alike wall begin the southward migration, thus the annual cycle is completed.

Each aspect of the behaviour of the black-headed gull can be related to the ecological problems that face the bird. In finding a mate of the proper species and sex, the ability to move from the winter feeding grounds to the colonial nesting area in which the gull was born or had already nested is critically important. Black-headed gull colonies are limited in number and are dispersed. By navigating directly to an established colony, the gull arrives at an area; (1) where reproduction should be possible given past performance at the site, and (2) where there are likely to be many reproductively active members of the opposite sex and proper species.

However, gull colonies are often composed of more than one species. The qualities of a location that make it attractive to black-hacked gulls also make it attractive to herring gulls, lesser black-backed gulls, and others. The potential problem of confusion in species identification has been overcome in part by the evolution of visual differences between species a well as species-specific long calls. This call is not only very loud, carrying a long distance to females far away, but it is qualitatively different from that of other species, although the basic pattern is somewhat similar. Black-headed gull females respond exclusively to males of their own species, avoiding or ignoring males of other species calling in the same general area
ignoring males of other species calling in the same general area.
Moreover, a male giving the long call is not only announcing

his species membership, but is identifying himself as a competent, reproductively mature individual capable of holding a territory in competition with other males.

When a reproductively mature male and female black-headed gull meet one another on a courtship territory, they do not copulate immediately. On the one hand, the female, because of her energy investment in reproduction, is under selection pressure to exercise caution in the choice of a mate. On the other hand, the male is primed to be aggressive, a highly adaptive trait if he is to hold a pairing territory. However, these aggressive tendencies are antagonistic to the sexual tendencies essential for mating. The male's reluctance to tolerate intruders in his area is overcome by a prolonged courtship that gradually reduces male aggression toward a specific female and develops sexual motivation.

A drawn-out courtship also leads to the formation of a stable pair bond. Pair formation is basic to the gull's strategy for maximizing its reproductive output. Gulls and many other animals have evolved elaborate adaptations for taking care of a relatively few fertilized eggs. These animals invest their time and energy in elevating the probability that a high percentage of the few offspring they do produce will survive to breed. The pair bond is, in effect, a cooperative agreement by both mates to contribute to nest building, incubation, feeding, and the defense of their clutch and brood. The male's contribution will greatly affect the female's reproductive success, for she has a great deal to lose by mating with a male that eventually deserts her in favour of another. By making copulation contingent upon a prolonged courtship, the female may make desertion a poor alternative for a male. By the time copulation takes place, unpaired females may not be available, thus making assistance of his mate the optimal strategy for a male. Moreover, by copulating with a male only after he has fed her often, a female can determine whether her mate is a competent food gatherer and therefore capable of being a good parent for her offspring.

In addition to pair bonding, the adult birds exhibit a host of other anti-predator adaptations that benefit their young. The fact that the gulls nest colonially is indicative of the strong selection pressures exerted by predators. Nesting in a tightly packed group creates numerous problems. There is less food nearby, forcing aduls to make long foraging trips. There is a greater possibility of interference in nesting from conspecifics leading to the evolution of vigorously defended pairing territories. This, in turn, favours highly aggressive males whose very aggressivity may interfere with pair formation, and so on. The severe disadvantages associated with colonial breeding are outweighed almost entirely by the benefits gulls enjoy from reduced predation. By nesting together, one bird's alarm call alerts dozens or hundreds of other gulls. The probability of a predator approaching an individual nest unnoticed is much reduced. Furthermore, coloniality permits group harassment of a predator, distracting and confusing it much more effectively than if a single pair were forced to defend their nest alone.

Another indication of the importance of predator pressure on the evolution of black-headed gull behaviour is the removal of eggshells by the parents. Moreover, the attraction call and the ability of adult birds to learn to recognize their mobile offspring permit the gulls to locate their progency after the chicks have scattered and hidden themselves in response to an alarm call.

The Kittiwake

Unlike almost every other gull species, the kittiwake nests exclusively on tiny cliff ledges on inaccessible to predators and thought to have evolved from ground nesting gulls as a result of predation pressure. There is every indication that its ancestors once nested on the ground. Kittiwakes retain some features of groundneating gulls, like the partial cryptic colouration of their eggs. The eggs of ground nesting birds usually are well camouflaged as a protection from predator but this can not be the function of egg colouration in Kittiwakes because each nest is marked couspicuonaly by white droppings. (Although mottled eggshells no longer have on antipredator function for kittiwakes, perhaps they occur because the female disposes of pigmented waste products by incorporating them in the eggshell).

FIG. 1.2: Kittiwake Gull

Culler studies a breeding colony of Kittiwake nesting on a very narrow cliff ledges. She observed that the eggs are free from the ground predator like rats. Badgers and foxes cannot climb the cliffs. Large predatory birds such as crows, herring gulls, and hawks have difficulty maneuvering onto the tiny strips of rock ledge on which kittiwake nests are built because of the powerful wind currents that swirl over the face of ocean cliffs. Kittiwakes can manage thanks to the evolution of structural adaptations that permit them to exploit these areas. The kittiwake is a small, delicate and lightly built gull with very strongly developed claws and toe musculature that enable it to cling to ledges upon which another gull would never be able to alight. The net effect of the kittiwake's choice of a nesting habitat is to reduce the effects of selection exerted by predators. Other radical changes in selection are all related to the properties of cliff ledges (they are narrow and therefore dangerous for young kittiwakes, for example) and the fact that suitable nest sites on a cliff face are not numerous. These changes have had significant effects on the evolution of kittiwake behaviour.

When a male returns in the spring to a cliff face on which he was born or has nested, he attempts to establish a territory on a ledge. This may consist of a strip of rock about 2 feet long and 6 inches wide. The male always courts the female and builds a nest at this site, unlike black-headed gulls, which often move from pairing territories to nesting territories. The kittiwake does not announce space possession with a long call, but instead employs a choking display, in which it tilts up and down while pointing its beak at the ledge and uttering a soft call. This behaviour is very reminiscent of placing material on a developing nest. Males mostly avoid chocking males, whereas unpaired females, upon their arrival at the breeding site (this occurs after the males' arrival), are attracted to displaying individuals. When a male does approach another male on its territory, the owner will engage in chocking. If the intruder does not fly off, the the birds may grapple, each holding the other by its beak, twisting and pulling in an attempt to throw the opponent off the narrow ledge. The intruder almost always loses.

When a female approaches a male and alights on the ledge, she turns her head and bides her beak in the feathers of her breast. A female may initially be driven from the ledge anyway, but her persistence gradually causes the male to tolerate her presence. Eventually the female performs head-tossing movements that cause the male to regurgitate food. Unlike ground-nesting species, the male does not simply spew his offering on the ground in front of the female; instead, he holds it in his throat and beak and the female takes the food directly from him.

Nest building involves the collection of mud and grasses from open pastures. The birds tend to fly in groups to spots were they can collect these materials. In addition, whenever, a kittiwake has a chance, it will steal a neighbour's nest-construction supplies; this never happens in ground-nesting gulls. Moreover, the nest itself, far from being a casual collection of grasses, is an elaborate affair with a stable mud or seaweed base (formed by a specific Trampling behaviour) and a deep grassy cup.

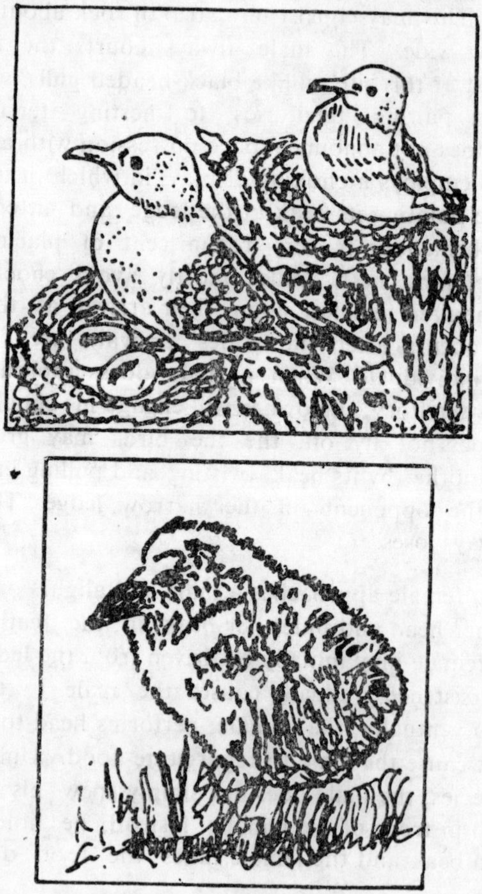

FIG. 1.3: Chicks of Kittiwake and of a Ground-nesting Gull

When the young hatch out, they are a highly conspicuous silvery white and not the usual camouflaged mottled brown. The eggshells are left by the nest. In addition, the parents rarely sound the alarm call and, when they do, there is no group mobbing of the predator that elicited the call. The young birds stay in the nest until they are mature enough to leap off the cliff, sailing down to the water dozens or hundreds of feet below. Because the juvenile kittiwakes always remain in the nest prior to fledging, there is no feeding attraction call given by the adults. And when an adult neighbour threatens a nearby nestling, the young bird odes not run away (as ground-nesting juveniles do), but instead hides its beak. As it bends its head down, the bird exposes a black neck band thought to provide another signal of submission in addition to beak hiding. Juvenile ground nesters lack a similar plumage pattern.

By comparing the behaviour and ecology of the kittiwake with that of the ground-nesting species, one can link almost all the differences between them to the evolutionary consequences of nesting on cliffs. We shall now turn to the behaviour of the green heron, a bird not closely related to the gulls, and attempt to relate behavioural similarities between these species to shared ecological pressures.

The Ecology of Green Heron

This is a rather small heron that nests either in isolated pairs along streams or in loose colonies in marshes. Its reproductive behaviour has been studied carefully by Andrew Meyerriecks. In this males in this migratory population arrive in the spring before females, and each actively defends a nest tree (usually one that already holds the remains of last year's nest). The male has a loud and highly distinctive call (skow), which it utters repeatedly while on its tree or flying around in the area about the tree. Other males tend to avoid "skowing" males. Should a male enter another male's territory, the owner may fly out to drive it off. If the intruder manages to alight on the nest tree, the territory owner will certainly challenge him. This may ultimately take the form of the full-forward display. The

TABLE 1.1

Selecsion Pressures and the Kittiwake's Key Adaptations

Altered selection Pressure	*Behavioural adaptations*	*Structural adaptations*
Absence of predators	Alarm call rarely given No group mobbing of a predator	Young not
camouflaged	No removal off egg shells	
Limited number of nest sites	Territory establishment at future nest site only Choking display signals territory ownership Territory very small	
Absence of other gull species in colony	No species-specific long call given to announce territory ownership	
Absence of nest materials at nesting sites	Group collection of material (when birds are expzosed to predators) Stealing of nest material common	
Narrowness of ledge	*In adults*: Aggression directed toward beak rather than upright-forward-charge sequence Beak hiding by females Nest with mud base, deep cup Mate and and young fed from throat; food not dropped	*In adults*: Small size, strong toes and claws
	In juueniles: Nest never left during development Beak hiding, not running when threatened	*In juveniles*: Black neck

bird is oriented horizontally with its feathers fully erected, tail flipping up and down, mouth open, exposing a brilliant red lining. With its deep orange eyes bulging, the heron may lunge at its opponent, waving its wings and uttering a harsh raaah call. As one might expect, the effect of this display is to drive the other male away in most cases.

Females are attracted to calling males. However, their attempts to approach are rebuffed initially by forward-displaying

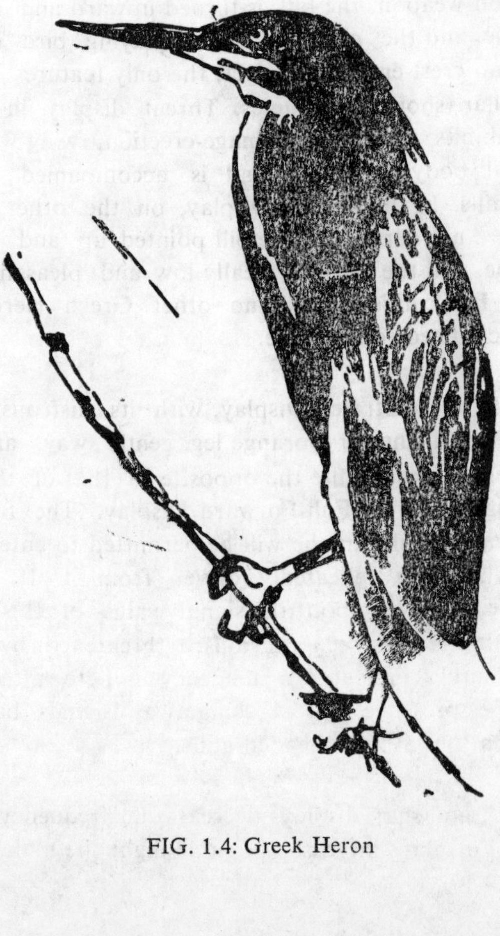

FIG. 1.4: Greek Heron

males. Despite this, females persist and male behaviour changes over time. The onset of courtship is signaled by the male's stretch display and snap display. The snap display is somewhat similar to the forward display, but the bird's beak is pointed down and its mandibles snap together. This display, which may have its origins in behaviour patterns used to gather nest material, is similar to choking behaviour in gulls and occurs at the start of courtship in green herons, as is true for kittiwakes but not for black-headed gulls. In both kittiwake and green heron, territoriality and courtship are focused entirely on the future nest site. The stretch display has been described in an especially vivid way by meyerriecks. In the Stretch Display the chief heron weapon, the bill, is turned upward and away from the female, and the neck of the displaying bird is exposed. There is no crest-erection; in fact, the only features erected are the scapular (shoulder) plumes. Threat display in the Green Heron exhibits extensive plumage-erection, wing-waving, low, horizontal body positions, and is accompanied by harsh, rasping calls. In the Stretch Display, on the other hand, the feathers. . . are depressed, the bill pointed up and away from the female, and the call is typically low and pleasant sounding (aaroo). This call is found in no other Green Heron display, and it is confined to the male.

A male in the Stretch Display, with its lustrous plumages, dark lores, brilliant coral-orange leg, gentle sways, and gurgling calls, offers a signal quite the opposite to that of the bristling, rasping fighter of the Full-Forward Display. The female must have a clear signal that she will be permitted to enter the nest after she has been repeatedly driven from it. It is not too difficult to speculate about the signal value of this behaviour to a female that has been previously threatened by the same male. It surely indicates a tendency not to attack and to tolerate her presence, just as facing away signals basically the same intention in black-headed gulls.

Stretch and snap displays decrease in frequency somewhat as mutual displays increase. In circle flight the male and female

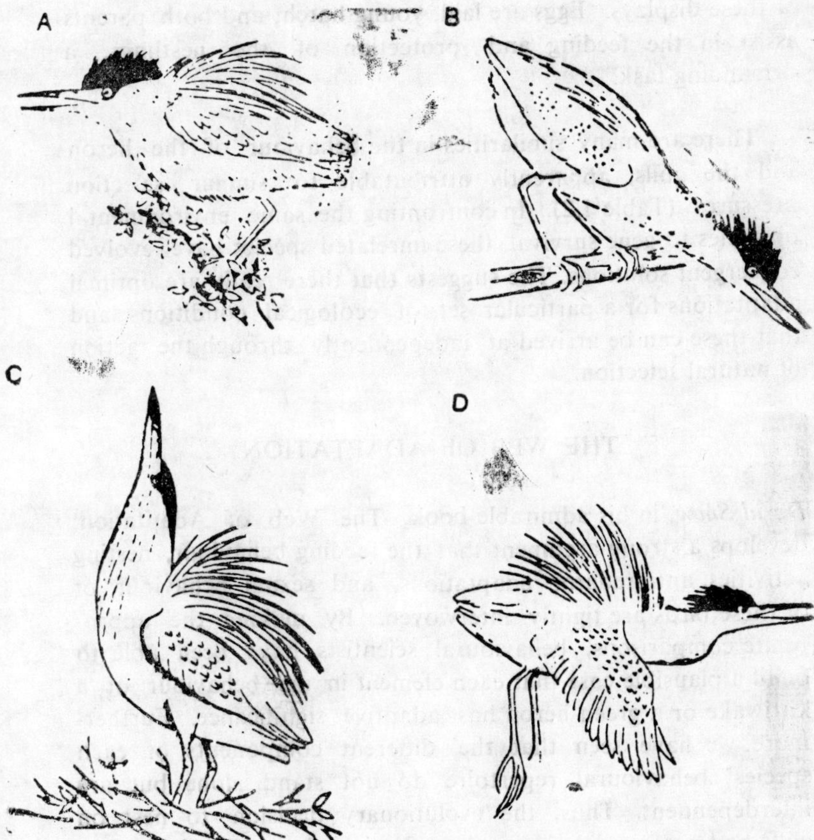

FIG. 1.5: Four Displays of the Green Heron. (A) The aggressive fullforward display. (B) The snap display given by courting males. (C) The stretch display; (D) The flap-flight display.

pair fly around the territory together, flap flight is similar, but with more exaggerated wingbeasts, presumably indicating a greater intensity of performance of this display. Mutual displays are followed by contact displays. The male and female may stroke their beaks together (billing) or preen the feathers on each other's necks (nibbling). These actions signal a readiness to accept contact between one another and as such are the foundation for copulation, which occurs shortly after the onset of these displays. Eggs are laid, young hatch, and both parents assist in the feeding and protection of the nestlings, a demanding task.

There are many similarities in the behaviour of the heron and the gulls, apparently attributable to similar selection pressures (Table 1.2). In confronting the same environmental obstacles to gene survival, these unrelated species have evolved convergent solutions. The suggests that there really are optimal adaptations for a particular set of ecological conditions and that these can be arrived at independently through the action of natural selection.

THE WEB OF ADAPTATION

David Snow, in his admirable book, The Web of Adaptation, develops a strong argument that the feeding behaviour, nesting activities, antipredator adaptations, and sexual behaviour of of these birds are tightly interwoven. By making the appropriate comparisons, behavioural scientists have been able to build a plausible case that each element in the behaviour of a kittiwake or a green heron has adaptive significance. Furthermore, we have seen that the different components of each species' behavioural repertoire do not stand alone but are interdependent. Thus, the evolutionary decision to nest on cliffs had profound consequences, not just for nest site selection by the kittiwake but also for the behaviour pattern involved in nest building, coping with predators, courtship and aggression.

The Contingas

Contingas tend to be fruit eaters, an unusual diet for birds in general. The kind of bill that is an efficient device for engulfing and swallowing whole fruits is not well suited for the construction of elaborate nests that require the precise manipulation of grasses and twigs. This may have restricted the nest-building options of fruit-eating cotingas, which as a group build very simple nests either on limbs of trees or in holes in trees. The selection of a nest site and the nature of the nest are also influenced heavily by predation pressure, which is intense in the tropics and favours inconspicuous or concealed nests. The bellbird, for example, builds a ridiculously reduced nest consisting of a very few twigs and tendrils balanced on a limb. The nest is almost impossible for a human observer to detect; despite intensive efforts to finds the nests of these birds, many of which are of robin size or larger, remarkably few have been found.

As a result, a female cotinga can meet her own food requirements in a few hours, leaving her free to incubate or gather food for her offspring the rest of the day. The fact that each female rears only one or at most two chicks, partly on superabundant fruits, contributes to her ability to take care of the nestlings single-handedly.

Diet and predator pressure together make it difficult for male cotingas of the limb-nesting species to contribute to the success of their progeny through parental care. Instead, they devote full time to highly competitive courtship activities in efforts to inseminate as many females as possible. In their courtship displays, the males employ their distinctively bright plumage, which contains red and yellow pigments sequestered from the fruits they have eaten.

The fruit-eating, cotingas that nest in holes, the tityras, are subject to a somewhat different set of selection pressures. They have a slightly larger clutch, presumably because the added eggs do not increase the conspicuousness of the nest, which is hidden in the nest chamber. This increases the value of male parental care slightly because there are more progeny to rear and thus greater food demands to satisfy. Male tityras do form pair bonds and contribute to the care of the young. The primary value of their contribution, however, may be to defend the nest site. Tree holes in the tropics and elsewhere are scarce and valuable resources for which there is intense competition.

TABLE 1.2

Behavioural Similarities Among Green Heron, Kittiwake and Black Headed Gull

Trait	*Green heron*	*Blacheaded gull*	*Kittiwake wak*	*Interpretation*
Males arrive at nesting area before females	X	X	X	This reflects the importance of territory ownership by a male if he is to acquire a mate; early males get territories
Territory established only at future nest site	X		X	Nest sites are limited for the heron and kittiwake; this favours males of these species that quickly acquire a nesting area
Males give loud call to announce territory ownership	X	X		Green herons and blackheaded gulls are under selection to attract distant females in areas with other related species; species recognition is a lesser problem for the kittiwake
Females are attracted to territory owners only	X	X	X	Females of all three species are under selection to mate with competent males and to choose nesting areas that increase the chances of reproductive success

Males are likely to attack females initially	X	X	X	Selection for a high degree of territorial aggressiveness in males leads to a low threshold for attacking any intruder into a male's territory
A forward display given with aggressive signal value	X	X	X	Facing and learning toward an opponent with the beak ready to strike (or beak wrestly—kittiwake) have an unambiguous message for the opponent
An early courtship display derived from nest building	X		X	The nest and territory are synonymous for the heron and kittiwake, but not for the blackheaded gull; a movement associated with nest building also signals a male's territory ownership to a potential female mate
Courtship display sequence: first, individual displays; then mutual displays; finally, contact displays.	X	X	X	Selection favours different displays at different stages of pair bonding: first the birds convey their uncertain tolerance of each other, then their growing willingness to cooperate, and finally acceptance of contact and copulation; this reduces male aggression, synchronizes male and female reproductive physiology, and provides the basis for mutual cooperation in rearing offspring.

A male tityra that left his mate after copulating with her might rarely produce surviving offspring because the female would be ousted from her nest by competing birds that attacked as a couple. Unlike their limb-nesting relatives, male tityras are not dramatically plumaged with distinctive colour patterns, but instead are similar in appearance to the famale of their species.

Finally, there are a number of cotingas, the becards, that are highly insectivorous. The becard builds an elaborate domed nest at the tip of a branch in a high tree. The male forms a pair bond with his mate and invests heavily in the care of the three or four offspring they produce. Capturable insects are generally in short supply in most environments and tend to be much more evenly distributed than fruits. Furthermore, an insect gains no advantage by being eaten (unlike a fruit, which uses its consumer as a dispersal agent for its seed or seeds). Insects have evolved a host of adapations that make their detection and capture difficult. By helping his mate provide hard-to-get food for the progeny, a male becard may make a valuable contribution to his progeny. He thus, improves his fitness more by being a good parent than by attempting to mate with more females.

The Consequences of Fruit Eating

Differences in diet and most nest site selection appear to have had profound effects on the evolution of all aspects of the behaviour of cotingas. Closely related species subjected to different selection pressures have diverged sharply in behaviour, just as kittiwakes differ behaviourally from their ground-nesting relatives. Conversely, distantly related species have converged in their behavioural adaptations in response to convergent selection pressures. For example, the manakins, although they belong to a different family of birds from the cotingas, have a diet that consists largely of fruit. The evolutionary consequences of this diet are much the same for manakins as for cotingas (Table 1.3). The same holds true for fruit-eating animals that are even more distantly related to cotingas than manakins. The African bat Hypsignathus is a prime example of a mammalian species dependent upon scattered fruiting trees in the tropics. Its behaviour shows many similarities to that of the fruit-eating birds (Table 1.3).

TABLE 1.3

Convergent Evolution in the Behaviour of the Fruit-Eating Cotingas, the Manakins, and a Fruit-Eating Bat, Hypsignathus

Shared Elements of Behaviour

1. Diet is dominated by fruit
2. Male parental care has little potential value because, for the birds, food is abundant, clutch size is small, and males cannot protect the nest or young against most predators; for the bat, food is also abundant, males lack milk to feed their young, and the bat's large size makes it relatively free from predation.
3. Care of the young is done exclusively by females.
4. Sexual dimorphism in appearance and behaviour is highly developed in all three groups.
5. In the mating system of these species:
   (a) Males gather in display groups.
   (b) Males defend small perch territories in he larger display arena.
   (c) Females inspect several males before choosing a mate.
   (d) There is great variation in the mating success of males.

The erratic fruiting pattern of figs and other favoured trees draws large numbers of individuals to small areas to harvest the fruit during the short time it is available. This may make it extremely difficult for any one male to monopolize a female or group of females and prevent other males from copulating with them. (The more males present, the greater the cost of defending the females). The society of chimpanzees is apparently highly fluid (almost chaotic), with individuals changing group membership frequently. When a female is in estrus she may form a consort relationship with a male, but often she simply accepts all comes. Although males do not fight for access to the female (again perhaps because of the energetic costs and risk associated with fighting a large number of other males), they nevertheless may compete in other ways. Males possess well-developed courtship displays to induce a female to copulate. They mate rapidly, passing their sperm to the female quickly to avoid interruption from other males. Moreover, they have very large testes, six times the size of those of the gorilla. Because females may mate a number of times, males may engage in sperm competition, trying to improve the odds that their sperm will fertilize the female's eggs simply by providing more gametes than a competitor.

The heariest primates, the gorillas feed on more evenly distributed lush forest vegetation and are much less dependent upon fruit trees as sources of food Large groups of individuals do not form at small feeding sites. As a result of the generally dispersed population, some males can economically and effectively defend a small group of mature females from the relatively few competitor males they will encounter in their leisurely travels through the tropical forest. Male gorillas are one male monopolizing mates with several female and do compete for ownership of a harem, but once established a resident male has no competition for access to the receptive females in his band. There is little courtship, and copulation is relaxed and prolonged (lasting up to 15 minutes) in contrast to hurried chimpanzee mating which is having a breeding system where several males copulate with each oestrus female. If this argument is correct, the distribution of food resources may have many correlated and predictable effects on elements of behaviour that at first glance would seem to have nothing to do with feeding behaviour. All the behavioural adaptations of a species are interrelated and, therefore, influence and constrain each other.

Cost Ratio of a Behavioural Trait

Predictions about the correlation between ecological conditions and a behavioural trait implicitly suggest that the benefits of the trait exceed its costs in specified circumstances. If those conditions are changed, then the benefit/cost ration will be altered (where benefits and costs are measured ultimately in gain or loss of inclusive fitness. By definition, natural selection will lead to the elimination of traits from populations in which the cost of the trait exceeds its benefit. If we can (1) identify the benefits and costs associated with a trait, and (2) determine how an ecological variable will affect the benefits and costs, we can make formal predictions about the relation between the variable and the trait. For example, males of many species (*e.g.*, gulls, herons, becards and gorrillas) remain with their mates for sometime after copulating with them. Others (*e.g.*, bellbirds, manakins and chimpanzees) do not. By observing these species, one can generate a list of plausible benefits and costs of the "male remains in association with mated female(s)" trait. The benefits of the trait will tend to exceed the costs to the extent that the environment of a species either elevates the benefits or decreases the costs. The key variables that will determine the benefit/cost ratio of this behaviour are the abundance and distribution of receptive females (potential mates) and other males (potential competitors). These variables, in turn, are influenced by such things as the distribuion of food resources or nesting areas. They will determine; (1) how many females a male can expect to fertilize successfully either by guarding a mate or by searching for new mates, and (2) the costs of guarding a mate from other males.

Many different combinations of costs and benefits can generate similar benefit/cost ratios. For example, males may remain near their mates eirher if the costs of defense are relatively low and the gains moderate (as in the case of the gorilla, where male competitors are dispersed and a male can simultaneously defend a few female mates) or if the costs are moderate but the gains very high (as in the case of territorial kittiwakes and tityras, where a male helps his mate defend a rare suitable nest site that is essential for reproductive success).

Some predictions that can be derived from our benefit/cost analysis are that a male will tend to remain near his mate if:

TABLE 1.4

Analysis of the Evolutionary Benefits and Costs of a Behavioural Trait: A Benefit Raises the Inclusive Fitness of an Individual; A Cost Lowers it

Trait male remains near mates(s) after copulating with the female(s)

1. *Possible benefit*: If the male is near his mate, he can prevent her from copulating with other males and thereby improve the chances that his sperm alone will fertilize any of her eggs that are fertilized.

   Ecological variables: Factors that affect the probability that a female will mate more than once in a mating cycle. *Example*, the distribution of food resources will affect the clumping of males and females and so affect a female's opportunities for multiple mating.

2. *Possible benefit*: If a male is near his mate and her offspring, he can identify his probable offspring and direct parental care to them.

   Ecological variables: Factors that affect the value of male parental care. *Example*: the intensity of predation may affect the optimal clutch size for a bird; the smaller the clutch, the less valuable the assistance of the male in providing food for the young.

3. *Possible cost*: If a male remains near his mate, his ability to locate additional females may be restricted.

   Ecological variables: Factors that influence the temporal distribution of receptive females. *Example*, if the breeding season is protracted, receptive females may be found over a long period, and the male may gain more by leaving a mate to search for additional receptive females.

4. *Possible cost*: If a male remains near his mate, he may have to repel other males from her if he is to enjoy benefits 1 and 2. Defense of a male entails an expenditure of energy and the risk of injury.

   Ecological variables: Factors that influence the capacity of males to injure one another. *Example*, in a long-lived species there may be considerable variation in the size distribution of males, with some males much smaller than others and, therefore, more vulnerable to injury by larger males.

1. There is high probability that if he leaves she will copulate with another male and use the competitor's sperm to fertilize her eggs (*e.g.*, langurs);
2. The male can provide parental care for his offspring that greatly improves their chances of survival (*e.g.*, the tityras);
3. The probability of finding another receptive female is very low after the male has found and courted his first mate (*e.g.*, gulls);
4. Male competitors are spatially dispersed so that a male may defend his mate(s) with relatively low energetic costs and reduced risk of injury in defense (*e.g.*, gorillas).

Ecological Predictions

These predictions have been developed primarily by making comparisons among a limited number of birds and a few mammals. They can be tested by analyzing the behaviour of other species. An especially useful test of ecological generalizations is to find a single species whose members behave differently under different conditions. One can then ask whether the behavioural differences are linked to ecological factors in the predicted manner. This avoids a problem inherent in comparing different species where one can never be completely certain that phylogenetic inertia is not responsible for the failure of a species to evolve the predicted optimal solution to a shared ecological problem. Behavioural variation within a species is far more common than usually appreciated. For example, consider the intraspecific variation in guarding behaviour by male bank swallows. Pairs of these birds nest in dense colonies in sand banks. Females are often pursued and sometimes knocked to the ground and mated by males not bounded to them. This increases the value to a male of guarding his partner. At times the male will leave the burrow when the female leaves and fly right behind her during the entire time she is away from the nest. At other times, however, males are not nearly so solicitous of their mates and will leave them to pursue other females. Sometimes a group of pursuers can separate a male from his mate and then copulate with the female. Close continual guarding of a mate occurs only during the period just before and during egg laying (when a female's eggs may be fertilized). At this time the benefit of male guarding increases dramatically (because if he fails to protect his spouse she is almost certain to copulate with other males whose sperm may fertilize her eggs). The benefits of close guarding during these critical days outweigh the costs (the time and energy expended in guarding flights and the loss of opportunities for matings outside the pair bond). But once this period is past, the benefit/cost ratio falls and the males sometimes attempt to locate other females that appear to be in the egg-laying process. This behaviour supports the hypothesis that male guarding is related to the probability that the eggs of a mate will be fertilized by a competitor if the mate is left unprotected.

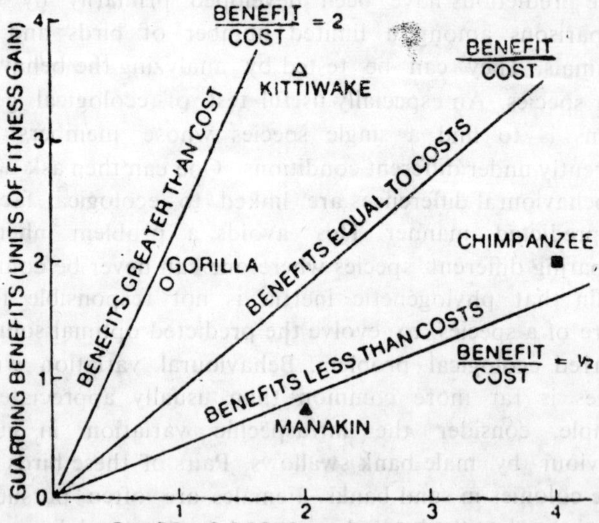

FIG. 1.6: Costs and Benefits of mate guarding by males. Different combinations of costs and benefits can yield similar benefit/cost ratios. The exact placement of the four species is speculative.

The currently more commonly employed test of the relation between a behaviour pattern and an ecological variable is to determine if a generalization developed for one group of animals holds for another group. Thus, if male birds guard their mates under some ecological conditions, then male frogs or male insects ought to do the same it these conditions really do provide the selection pressure for the evolution of male guarding. To take single example, males of many species of damselflies regularly remain near their mates following copulation. In these species, females are receptive only about the time they are going to lay their eggs. Males wait for females at oviposition sites (typically patches of submerged aquatic vegetation). The limited distribution of acceptable sites concentrates males spatially and increases the costs of guarding a female after

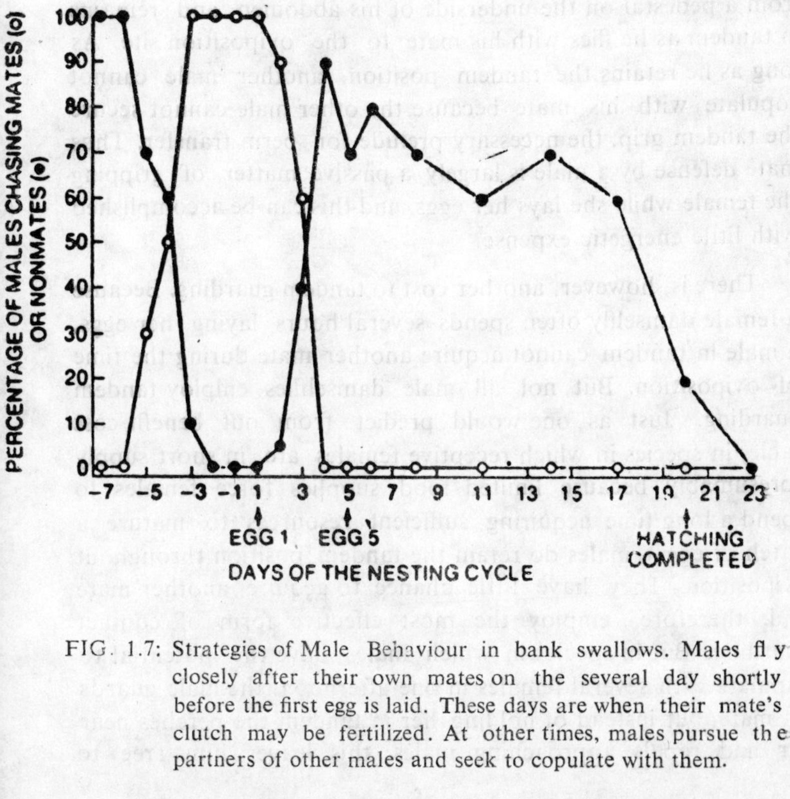

FIG. 1.7: Strategies of Male Behaviour in bank swallows. Males fly closely after their own mates on the several day shortly before the first egg is laid. These days are when their mate's clutch may be fertilized. At other times, males pursue the partners of other males and seek to copulate with them.

mating with her. However, a female damselfly may mate with more than one male if she is not guarded. The evidence suggests that when multiple mating occurs in insects, the last male to copulate before the eggs are laid usually fertilizes most or all of that female's eggs. Thus, the benefit of guarding a mate is extremely high because if she is not defended from other males there is little likelihood that a male will gain genetically from a copulation. Moreover, the costs of mate guarding are moderated by two factors. First, the body design of a damselfly is such that one male cannot easily inflict a serious injury on another; this somewhat lowers the risk-of-injury cost of guarding. In addition, when a male grasps a female prior to mating, he grips her on the anterior end of her thorax with special claspers at the tip of his abdomen. A male often retains this grip after the female has taken his sperm from a pedestal on the underside of his abdomen and remains in tandem as he flies with his mate to the oviposition site. As long as he retains the tandem position, another male cannot copulate with his mate because the other male cannot secure the tandem grip, the necessary prelude for sperm transfer. Thus mate defense by a male is largely a passive matter of gripping the female while she lays her eggs, and this can be accomplished with little energetic expense.

There is, however, another cost to tandem guarding. Because a female damselfly often spends several hours laying her eggs, a male in tandem cannot acquire another mate during the time of oviposition. But not all male damselflies employ tandem guarding. Just as one would predict from out benefit/cost table, in species in which receptive females are in short supply (presumably because limited food supplies force females to spend a long time acquiring sufficient resources to mature a batch of eggs), males do retain the tandem position throughout oviposition. They have little chance to acquire another mate and, therefore, employ the most effective form of contact guarding. But in species in which males have the potential to copulate with several females in one afternoon, the male guards his mate, but instead of holding her in tandem the perches near her and repels approaching males; this leaves him free to respond sexually to newly arrived females. Thus, not only do males of a group of insects guard their mates in a manner consistent with predictions derived from an analysis of vertebrate behaviour, but the kind of guarding employed by males of different damselflies appropriate for the different ecological conditions they experience.

Optimality Theory

By identifying the benefits and costs of a trait and by showing how these might vary as ecological parameters vary, one can make qualitative predictions about what animals should do under specified ecological conditions. However, an envolutionary approach does not predict merely that traits will remain in populations if their fitness benefits exceed their fitness costs. Rather, the prediction is that optimal traits will evolve—that is, traits whose benefits exceed their costs by a greater amount than any other alternative. If one could make quantitative determinations about the gains and losses associated with a trait, one could test whether in fact the trait was the optimal solution and did make the greatest possible contribution to gene-copying success. It is rare indeed that one can unequivocally measure the contribution made to inclusive fitness by any trait. In practice, optimality theory is usually applied to resource-gathering behaviour, with the assumption that the individual that acquires the resource most efficiently (*i.e.*, at the highest rate and with least expenditure of energy) will enjoy higher fitness than less efficient individuals. It is assumed, for example, that animals that can gather food at the highest rate will outcompete less successful ones in times of food shortages. At all other periods, efficient foragers should continue to enjoy an advantage because they will have more time for other important activities if they can satisfy their food requirements in as short a time as possible. These attributes should, the argument goes, be correlated with higher fitness.

To test whether an animal really does forage as efficiently as possible or move about as economically as possible or court a female as effectively as possible, one first lists the advantages and disadvantages of the components of the behaviour. Then one must translate the costs and benefits into the same units (fitness gains and losses ideally, but in practice usually calories gained or lost or females gained or missed). Then an equation is constructed and solved to maximize the benefit/cost ratio in order to determine what an animal should do in order to maximize such things as net energy gained per unit time or number of fertilizations accomplished per unit time. The actual behaviour of an animal can then be compared with the predicted optimum response. Although the mathematics of this approach are often intimidating, optimality theory nevertheless offers the most rigorous test of hypotheses about the match among behaviour, ecology, and inclusive fitness.

ECOLOGY AND BEHAVIOUR OF THE DESERT LIZARD APOROSAURA ANCHIETAE

Of the many desert lizards which have been studied few have been found to have specific physiological adaptations to the desert environment. The majority escape unfavourable conditions by well-defined adaptive behaviour. Escape from thermal stress and desiccation by behavioural means usually involves hiding beneath sand or rocks. This behaviour has the disadvantage that severely limits opportunities for feeding. This is particularly true of the lizard Aporosaura anchietae found in the Namib desert in South-west Africa. It typically inhabits the slip face on the leeward side of sand dunes. The leeward side of the dure acts as a trap for wind-blown organic matter which congregates at the foot face, where Aporosaura feeds upon grass seeds and insects. The physiological and behavioural adaptations of this lizard have been described in some detail by Lauw & Holm.

The leeside of the dunes is virtually wind free and frequently faces the rising sun. Consequently, the surface temperatures rise very rapidly. When the surface temperature exceeds 30°C the lizard emerges from the sand and presses its ventral surface against the substrate. It takes a special 'dished' posture which achieves the maximum contact between its body and the substrate. By this behaviour the animal achieves a rapid rise in body temperature and is soon able to move about and forage on the dune face. As substrate temperature approaches 40°C the animal straightens its limbs in a characteristic 'stilt-walk', which raises the body as far as possible above the substrate. It periodically raises the diagonally opposite limbs while the base of the tail is used for support. At about 40°C the lizards dive beneath the soft sand to reach the cooler depths. The behaviour of Aporosaura thus appears to be very much dictated by its thermo-regulatory requirements. These generally result in a biphasic period of activity during which the animals is able to forage on the slip face. The work of Louw & Holm shows that these lizards are able to consume large amounts of food in a short period of time. After feeding in this manner, captive animals exhibit little or no interest in feeding and remain submerged beneath the sand for longer periods than usual. These workers also found that Aporosaura is able to store water in the digestive tract. It is capable of ingesting water amounting to 11% of its body weight within three minutes and can subsequently survive for more than eight weeks without further opportunity to drink. These lizards obtain water primarly through eating insects; though they also have occasional opportunity to drink condensed water from periodic advective sea fog, which is typical of the area.

When the lizards are under the surface of the sand, they are protected from climatic changes and from predation. However, they have no opportunity to obtain food or water. When weather conditions are favourable during the day, the animals emerge upon the slip face and feed upon detritus and arthropods. Here they suffer from a much increased rate of dehydration and are vulnerable to predation by the sidewinding adder (Bitis peringueyi), the black-backed jackal (Canis mesomelas), the chanting goshawk (Melierax musica) and the rock kestrel (Falco tinnunculus). The lizard's normal reaction to danger is a fast spring followed by emergency diving beneath the loose sand. It is capable of this rapid anti-predator response only when its body temperature is in the region of 30-40°C.

Below these temperatures the lizards are less mobile, and on cold days they generally remain below the surface of the sand.

On the basis of these observed behaviour patterns of Aporosaura, and of their probable survival value, we can make a guess about the optimum survival strategy for this species. When the animal is replete with food and water we would expected it to spend less time on the surface of the sand and thus avoid predation and dehydration. When it has a negative energy or water balance, it can avoid eventual death only by venturing on to the surface of the sand to search for food and water. In cases of extreme hunger or thirst, we would expect the animal to run a greater risk of predation by emerging from the sand at a lower temperature than normal. In such circumstances a period of specialised warming behaviour would be necessary. Similarly, we would expect the animal to run greater risks of dehydration by remaining on the surface at ever-increasing temperatures and indulging in bouts of specialised cooling behaviour.

We now have the problem of accounting for the mechanisms by which the animal achieve such appropriate behaviour in changing environmental circumstances. The traditional approach is to look for the sense organs through which information about the external environment, and about internal physiological state reaches the animal's brain. On the basis of work on other lizards, we have good reason to suppose that central thermoreceptors and dehydration monitors might exist in Aporosaura. The representation of such information in the brain is commonly thought of as some kind of drive, or potentiality, to perform appropriate behaviour. The drives compete, or interact, in such a way that the observed behaviour is controlled by the strongest drive.

Let us pretend that the animal merely has to make a decision to remain above, or below the surface. In accounting for the adaptiveness of lizard's behaviour, knowledge of the nature and strength of the relevant drives would not be sufficient. It is obvious that there must be some calibration of the drives, in terms of their relative importance. We would then have to specify how calibrated drives combine and interact. The necessity for both calibration and combination can be illustrated by reference to the stylised Aporosaura example.

In investigating calibration of the monitored internal states relevant to temperature and dehydration, we would expect to find that the calibrated drives were related to the survival value of the physiological states. For example, the probability of death resulting from thermal stress might well be a function of temperature and the probability of death by predation might also be function of temperature, for an animal on the surface, because the lizards are less mobile at lower temperatures. On this basis we might expect the calibration of the thermal drive. Similarly, we might expect the calibration of dehydration the tendency to surface increases with dehydration, thus increasing the animal's chances of finding water; but at extreme dehydration this advantage is offset by the increased loss of water that results from being on the surface and exposed to the sun and wind. We not have a calibrated tendency to surface, based on temperature and a similar tendency based on dehydration. The question is—how do these combine to produce a single tendency to surface?

In order to illustrate the difficulty of this aspect of the problem, Bateson & Hinde have calculated the consequences of assuming that the combination law is additive, and have separately calculated the consequences of assuming that the law is multiplicative. The additive rule implies that the isocline representing a particular tendency to surface is linear, while the multiplication rule implies that the isocline is a hyperbola. If we take these isoclines to represent the threshold tendency to surface (*i.e.*, the tendency which, on average, will be large enough to cause the animal to be above the surface), then we can represent the consequences of the hypotheses embodied in the two isoclines in terms of a map on the physiological coordinates. In other words, on the basis of the (hypothetica l) calibrations the additive hypothesis produces the map shown by a thick line and the multiplicative hypothesis produces the dashed one. Clearly the manner in which the calibrated drives are combined makes a large difference to the adaptiveness of the resulting behavioural strategies.

The problem with this approach is that it is difficult to foresee the day when the physiological analysis of behavioural control mechanisms will have advanced to such an extent that it will be possible to measure the way in which drives are calibrated, and combine to produce decision rules appropriate to the animal's way of life. It is equally difficult to see how traditional motivational analysis, with its view of behaviour as driven from within, can ever succeed in this task. The traditional approach to motivation is to attempt to specify the link in the causal chain between the animal's physiological state, its perception of the environment, its calibration of these factors in terms of their importance of survival, and its formulation of appropriate decisions, which are translated into behaviour. The relevant consequences of behaviour can be divided into those which are perceived by the animal, and alter its motivational state; and those which are not relevant to the behaviour of the animal in the short-term, but nevertheless are subject to natural selection. For example, an animal may eat a poisonous plant without perceiving the consequences as being any different from those of eating a non-poisonous plant. Even if the poison does not kill the animal, but merely makes it sick, ingestion of the plant may affect the animal's chances of survival by temporarily lowering its general efficiency, ability to avoid predators, etc. Naturally, we would expect that in the long term the animal would develop a suitable avoidance mechanism, such as that responsible for bait shyness in rats. However, we can never assume that there is always, or even mostly, a match between the consequences of behaviour as perceived by the animal, and those acted on by natural selection. Nevertheless, every behaviour sequence will cause changes in the internal state of the animal, that can be related to the survival value of that same behaviour sequence because every behaviour sequence has both of the types of consequence discussed above, and these are highly correlated. The correspondence between changes in the internal state of the animal, and the survival value of the behaviour responsible for some of those changes, forms the basis of a new approach to the analysis of motivational mechanism in terms of both form and function.

2

Environmental Behaviour in Insects

The study of organisms in relation to their environment is the modern science of ecology, which has developed in recent years from the earlier, essentially descriptive endeavour, natural history. This inclusion of this chapter under the general heading "Structure and Function" is consistent with Odum's (1971) comment that "it is more in keeping with the modern emphasis to define ecology as the structure and function of nature."

A number of general ecology texts are available Price (1975) deals specifically with insect ecology. Andrewartha and Birch (1973) trace the historical development of insect ecology.

A basic functional unit in ecology is the ecosystem, which is composed of a community of living (biotic) interacting organisms (plants, animals, and micro-organisms) and the nonliving (abiotic) components of their environment. Ecosystems vary in size and complexity, ranging anywhere from a woodland pond to the entire biosphere (that portion of our planet in which living forms exist). The biotic part of an ecosystem is made up of producers, consumers, and decomposers. Producers are green plants, which synthesize food (carbohydrates and other compounds) from carbon dioxide and water by utilizing energy derived from solar radiation (*i.e.*, photosynthesis). Consumers are those organisms, the vast majority of which are animals, that obtain energy and molecular building blocks by ingesting organic matter from plants (herbivores and detritivores; or from other consumers (carnivores and detritivores; Decomposers, mainly bacteria and fungi, break complex organic molecules down to simple inorganic nutrients usable by producers, facilitating the continuous cycling of inorganic nutrients throughout an ecosystem.

The Life-System Concept

Clark *et. al.*, (1967) introduce a particularly useful concept for our purpose, the concept of the life system. A life system is "that part of an ecosystem which determines the existence, abundance, and evolution of a particular population." In other words, the life system of an insect is that part of the environment (effective environment) that directly influences the fate of a given population plus the population itself. Obviously not all parts of an ecosystem necessarily directly influence a given animal population, although the situation becomes rather hazy when one considers indirect influences. The essential components of a life system are presented in Fig. 2.1. In view of the life system concept, an ecosystem becomes a series of interlocking life system.

The following sections deal with various aspects of the major components involved in the functioning of a life system, *i.e.*, population and effective environment.

Populations

Members of given insect species are typically separated into more or less discrete groups or populations. Or a population may be a group of individuals of the same species delineated, for purposes of study, by a biologist. Like individuals, populations have characteristics that can be measured and described: for example, genetic composition (the individuals of most

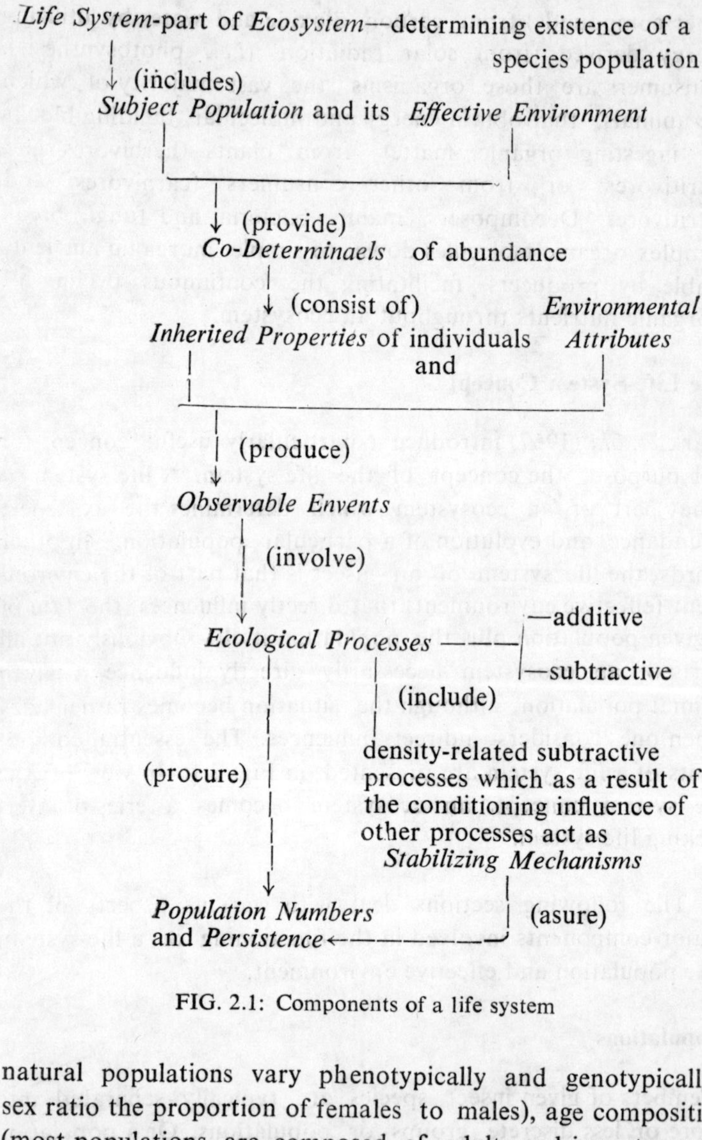

FIG. 2.1: Components of a life system

natural populations vary phenotypically and genotypically), sex ratio the proportion of females to males), age composition (most populations are composed of adults and immatures of varying ages), arrangement in space or dispersion (most natural populations tend to be composed of clumps of individuals due to irregularities in the distribution of food and shelter and the fact that individuals often attract one another), population size expressed as total number of individuals) population density (the number of individuals per unit area or volume) biomass (the total weight of a population), and dynamics changes in number and/or density over time). It has been long recognized that organisms have the capacity under ideal conditions to increase by a geometric progression (exponentially). Since this capacity for exponential increase (biotic potential) is never fully realized under natural circumstances, it is clear that there are factors in the environmental (predators, limited resources, etc.) that act to prevent such increases. The collective action of these factors has been called environmental resistance. In the broadest terms, one may view the dynamics of a given population as representing the outcome of the interaction between biotic potential and environmental resistence.

It is the goal of population studies to understand the composition and behaviour of populations and ultimately to be able to make predictions. Such information is essential to the development of a detailed understanding of evolution and further of paramount practical importance when we consider pest organisms.

Three basic approaches are used to study populations:

1. Analysis of population performance under controlled, but artificial, conditions in the laboratory.
2. Evaluation of populations in the field.
3. Development of theoretical, mathematical models that describe population dynamics.

Population studies in the laboratory are of necessity too simplified and only remotely representative of what must actually occur under the complexities characteristic of field conditions. On the other hand, the complexities of natural conditions make it difficult to access the action of particular environmental factors. Further, such studies must rely on sampling—studying a comparatively small number of individuals from a population with the hope that the characteristics exhibited by a simple accurately represent those of the whole population. Whether this hope is realized depends upon the adequacy of sampling. Sampling methods vary with insect species, habitat, and the kind of information to be derived from the sample. Many sophisticated sampling techniques bave been developed, and elaborate statistical analyses have been made easier by the use of computers. It is necessary that population studies be made over several generations. Theoretical studies are especially valuable in developing descriptive and predictive models of population performance; however, as with laboratory studies they tend to represent over simplifications and must rely on assumptions that may or may not apply in natural situations. Fortunately, modern computers allow evaluation of increasingly complex models. All three approaches to studying populations have some value and definite limitations. Increased understanding of populations wlll no doubt result from an amalgamation of ideas generated by all three approaches.

RELATIONSHIP BETWEEN ENVIRONMENTAL COMPONENTS AND POPULATIONS

Population Growth

This section establishes a simple, but useful model, which helps in thinking about population growth. Assume we are evaluating the repopulation of an area from which all members of a given species have been previously removed. Since populations under ideal conditions have the capacity for exponential increase, the first step is to consider an equation that describes exponential increase. Such increase in a population after time t is described by the equation $Nt=N_0e^{rt}$ where Nt=population size after time t; N_0=initial population size; e=base of natural logarithms (a constant; approximately 2.7183); and r=the instantaneous rate of increase per individual in the population, which is the difference between the instantaneous birth rate b (natality) and death rate d (mortality). Thus $r=b-d$. If b is greater than d, then the population is growing; if b is equal to d, the population is stable (r=0); if b is less than d, the population is decreasing in size. The graph depicted in Figure 9.2A shows the shape of a curve that would be generated by the exponential population growth equation if r remained at a constant positive value. The population size increases in an increasing rate.

Under natural conditions, r does not remain constant, being subject to strong environmental influence. For example, r may well decrease with increasing population density. Competition for food and shelter, rates of predation, and invasion by parasites and parasitoids are examples of environmental factors that may increase as population density increases. Looking again at the equation for exponential growth, but with r decreasing at a constant rate as population size increases. In this case, rate of population growth decreases gradually and approaches a point where it levels off. The population size at which the leveling off occurs is called the carrying capacity (K) of the enuironment. This curve is commonly called the logistic model of population growth.

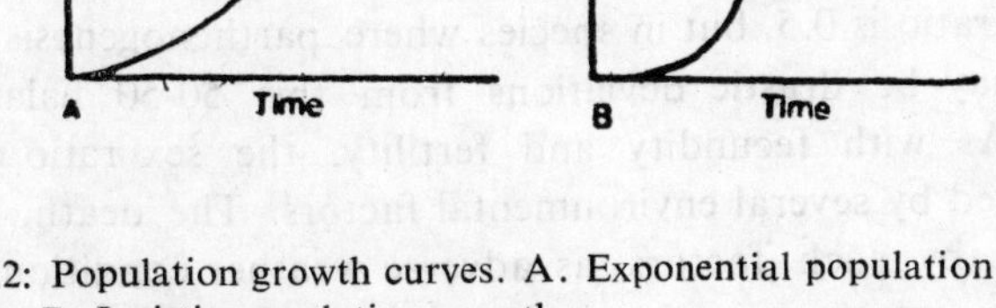

FIG. 2.2: Population growth curves. A. Exponential population growth. B. Jogistic population growth.

Although the logistic model may come close to describing population growth in some cases, it is severely limited because the assumptions upon which it is based are rarely true in reality. Among these assumptions are equality of reproductive potential among all individuals of the population, an even age distribution with a constant proportion of individuals breeding all the time, reproduction uninfluenced by climate or other factors, and no changes in carrying capacity.

In addition to birth and death rates, emigration and immigration also influence population growth. Birth rate, death rate, and emigration-immigration rates are themselves influenced by a variety of factors. For example, weather, in particular temperature, influences all these rates. In addition, it can be argued that any factor the exerts an influence on one of these rates influences the others as well. For example, an environmental change, say in temperature, might cause higher mortality among insects in one stage than in another or in insects of a certain age. If this age-specific mortality happened to occur in ovipositing females, one could say that both the birth and death rates were being affected. Among the major factors that influence birth rate are average fecundity of the females, average fertility of the females, and the sex ratio. Average fecundity represents the average number of offspring that would be produced by each female under ideal environmental conditions; average fertility represents the average number of offspring actually produced, the difference between fecundity and fertility being related to such environmental factors as quality and quantity of available food, weather components, and population density. The sex ratio is the faction of the totol population that is female. In most insects the sex ratio is 0.5, but in species where parthenogenesis occurs there may be drastic deviations from this 50-50 balance of sexes. As with fecundity and fertility, the sex ratio may be influenced by several environmental factors. The death rate is affected by such factors as adverse weather conditions, temperature extremes in particular; predators, parasites, and pathogens; accidents; low vitality; food shortage: and lack of adequate shelter. The carrying capacity (K) of the environment also varies with many factors (*e.g.*, availability of food, extent of predation).

More Reading the Life-System Concept

Several different models have been proposed to explain the numerical behaviour of insect populations. As previously mentioned, Clark *et al.* (1967) provide a useful integrating concept, *i.e.*, the "life-system". Our objective here is to elaborate somewhat on this concept. Two kinds of "ecological events" occur as a result of the interaction of the "co-determinants of abundance" (*i.e.*, the subject population" and its "effective environment"); primary and secondary. Primary events are those directly involved in the demographic equation explained above (*i.e.*, births, deaths, emigrations, and immigrations). Secondary events are those that exert an influence on the magnitude, extent, frequency, or duration of primary events (quality and quantity of available food, weather components, etc.). Primary and secondary ecological events may be additive or subtractive. Processes that act in a positive way on a population are additive, for example, immigration and weather conditions that reduce populations of natural enemies. The planting of a monoculture, such as corn, is an additive process relative to species, such as the European corn borer, that thrive on this crop, as well as to insects that are parasitoids or predators of the corn-eating species. Subtrartive processes (*e.g.*, emigration and adverse weather conditions) have a negative effect on a population, causing the death of individuals and/or a decrease in the number of progeny produced. The planting of corn would constitute a subtractive process for insects that were originally present, if they could not survive in a corn agro-ecosystem.

Additive and subtractive ecological events may be either density-independent or density-related (*i.e.*, the effects of the action of a given ecological process on a population may or may not be related to the density of that population). Subtractive density-related processes are of special interest since they may act, in some instances, as regulators of the level of abundance: that is, as a given population increases, a point is reached where further growth in inhibited by some factor.

Long-Term Numerical Changes

Over the long term, as populations interact with their effective environments, there are fluctuations in population size. Some populations fluctuate irregularly, apparently in response to changes in environmental components such as food supply and weather. Other populations tend to fluctuate regularly about a mean level of abundance. As mentioned above, this may be a reflection of the action of some subtractive density-related process that is serving to regulate the level of abundance. Insects with a reproductive period restricted to a particular time or times of year tend to display regular, seasonal population peaks associated with reproduction.

It has become more and more apparent that, in a given life system, certain "key" ecological processes with age-specific effects determine major population trends, while other processes exerts somewhat lesser effects. Elaborate "multifactor" studies have been carried out for a number of insect species. The objective of these studies has been to derive "life tables" that contain information as to population densities at different times during the life cycle and age-specific information regarding the "key" ecological processes that account for mortality. Such a life table is presented in Table. The development of a truly representative life table requires the sampling of a population over many generations. Another approach that has been used where preliminary studies have indicated a few key influences involved in a life system in the key-factor method. In this type of study, population density is measured at only one point in each generation. This approach requires much less sampling and enables investigators to concentrate on critical periods of time in the life when the "key" process or processes are active without spending much time on other periods.

Environmental Components

Various components of the environment that have been found to be significant in their influence on insect or other animal or

TABLE 2.1

Life Table for Second Generation of the Diamondback Moth on Early Cabbage

Age Interval	*Numbers Per 100 Plants*	*Causative Factors*	*Mortality Per 100 Plants*	*Per Cent*
Eggs	1580	Infertility	25	1.6
Larvae				
Period 1	1555	Rainfall	1299	77.1
Period 2	356	Rainfall	36	10.1
		Parasitism by *M. plutellae*	52	14.6
Period 3	268	Parasitism by *H. insularis*	69	25.7
Pupae	199	Parasitism by *D. plutellae*	92	46.2
Moths	107	Sex (49.5%00)	1	1.0
Females×2	106	Photoperiod	78	73.6
"Normal" females×2	28	Adult mortality	20	71.4
Generation totals			1572	99.5
Trend index [b]0.55 Stable number rate 99.1%				

[b]Trand index $= \frac{\text{eggs laid in new generation}}{\text{eggs laid in old generation}}$

[c]Stable number rate = the mortality rate necessary for the population in remain the same size from generation to generation. Since the percent mortality for the generation represented in this table was 99.5%, it is apparent that the population was declining.

plant populations are discussed in this section. Any combination, or all of these components, may at some time act to varying degrees upon a given population. When one of them approaches or exceeds the species-specific limit of tolerance, whether expressed in terms of survival, development, fecundity and so on, it become a limiting factor. That is, it becomes the environmental component that is of development, fecundity, and so on. In the field it is often very difficult to pinpoint a given environmental component as a limiting factor since, obviously, several components are likely to be interacting to produce the "effective environment" of any population. For the same reason, it is difficult to relate studies of environmental effects under the carefully controlled conditions of a laboratory to what actually occurs in the field. Another problem further complicates the picture. Environmental components, in particular temperature, moisture, and light, are not uniform throughout an ecosystem. For example, the temperature of one part of a plant may be quite different from that of another part. Thus, at least in many instance, to obtain a truly accurate picture, one must measure the temperature in specific parts of an ecosystem (*e.g.*, bottom side of a leaf). Realization of this problem has led to the development of the concept of "microenvironment", which implies the recognition of distinct horizontal and vertical (spatial) and temporal differences in environmental components within an ecosystem. Figure portrays Such differences in terms of relative humidity and water vapor pressure at different levels above the ground (vertical) and at different times during the 24-hour cycle (temporal).

Weather

Weather results from the combined action and influence of all the physical factors of the environment at any given time. It varies continually throughout days, weeks, months, and years and exerts an influence on insect abundance, longevity, rate of development, and so on, from one year or season to the next. Climate, on the other hand, is the everage course or condition of the weather in a locality over a period of several years. Weather changes rapidly, often violently, while climate tends to remain pretty much the same or change very slo ly over a period of many years. The main elements of weather are temperature, moisture, and light, although several other physical environmental factors are known or thought to exert a degree of influence on insects.

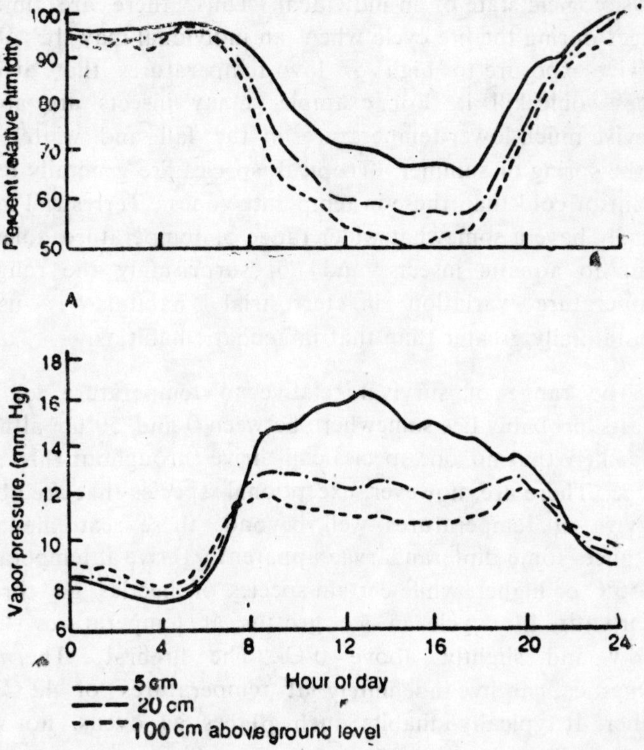

FIG. 2.3: Changes in relative humidity (A) and water vapor pressure (B) at different levels above the ground and at different times of day.

Temperature

Insects are basically poikilothermic—that is, their body temperature tends to be the same as ambient temperature. however, this does not necessarily mean that an insect's body temperature is always the same as that of the environment. For every insect species there is a fairly well-defined range of temperature within which it is able to survive, Exposure to temperatures above the high or below the low extremes of this range results in death. The range of tolerable temperatures varies from species to species, within a species, and with the physiological state of an individual. Thus, there are times or stages during the life cycle when an individual may be able to survive exposure to high or low temperatures that at other times would kill it. For example, many insects are able to sugvive much lower temperatures in the fall and winter than in the spring or summer. Tropical species are generally less to lerant of cold than these in temperate zones. Terrestrial insects usually have a somewhat wider range of temperature tolerance than do aquatic insects, and not surprisingly the range of temperature variation in terrestrial habitats is usually substantially greater than that in acuqatic habitats.

The range of survival relative to temperature for most insects probably lies somewhere between 0 and 50°C, although it is likely that no one species can thrive throughout this entire range. There are, however, exceptional species that are able to survive at temperatures well beyond these extremes. For instance, some dipteran larvae apparently thrive at temperatures of 55°C or higher, while certain species of beetles go through their entire life cycle in ice grottos at temperatures slightly below and slightly above 0°C. The firebrat, Thermobia domestica, can live indefinitely at temperatures of 42°C and higher. It typically inhabits such places as ovens, hot-water pipes, and simillar "hot" environments. If insects of a given species are exposed to temperature gradient, they will move until they reach the "preferred temperature," at which point they will tend to congregate. Near the upper and lower tolerable limits of temperature, insects become dormant. In the range between these limits, they are active.

The actual cause or causes of death at the limits of the temperature range are not clear. At the lower limit the submicroscopic structures of cells may be disrupted by the formation of ice crystals or the metabolic balance may be thrown off. At the upper limit protein denaturation, metabolic imbalance, disruption of ordered molecules (as, for example, in the wax layer of the epicuticle), and desiccation are likely to be involved. The phenomenon of insect cold-hardiness has been reviewed by Salt (1961) Asahina (1966), and briefly by Baust and Morrissey (1977). Downes (1965) discusses adaptations of insects in the Arctic. Many insects that become dormant in temperate regions are able to survive low temperatures for considerable periods of time. Most of these insects are dormant in a stage that is more cold-hardy than the preceding one. Some are capable of long exposure to low temperatures and display a certain amount of cold acclimation, but succumb to freezing of the body fluids. For example, Aedes aegypti larvae reared at 30°C are killed by exposure to -0.5°C for 17 hours, but survive such an exposure if preconditioned for 24 hours at 18 or 20°C.

Other insects are able to avoid freezing because they can withstand supercooling (*i.e.*, being cooled below the point of freezing, but without freezing actually taking place). Individual species vary as to the temperature to which they can supercool. Cryprotective compounds such as glycerol, sorbitol, and erythritol have been found in the tissues of overwintering insects. Salt (1959) showed that glycerol depresses the hemolymph freezing point of Bracon cephi larvae as much as 17.5°C and that this compound plays a key role in cold-hardiness. Glycerol begins to appear in the tissues in correlation with the advent of dormancy and usually disappears rapidly when the insects come out of this state. A number of insects (*e.g.*, many caterpillars) are able to survive the formation of ice within their bodies. Members of this group of "freezing-tolerant" insects have been found to possess higher concentrations of cryoprotective compounds in their hemolymph. Apparently these compounds act to prevent tissue destruction by influencing how and where ice crystals from as well as depressing the hemolymph freezing point. The lathal effects of high temperatures are difficult to study because of the influence of moisture changes. The interaction of moisture and temperature is considered later. It has, however, been established that lethal temperatures may very depending upon the temperature to which an insect or, more properly, a sample of a population of insects has been exposed previously. Such high temperature acclimation has been demonstrated in Calliphora and Phormia and others. Acclimation to both high and low temperatures may or may not occur in the same species. Like cold acclimation, high-temperature acclimation is likely to be of value to insects under natural circumstances. Since seasonal and daily high temperatures are usually preceded by a gradual transition from somewhat lower temperatures. Acclimation at both ends of an insect's tolerable range of temperatures may be looked upon as promoting the insect's survival against the effects of extreme daily and seasonal temperature fluctuations.

Temperature also affects the duration of life. For example, a clear relationship exists between the duration of life in male and female Drosophila subobscura and temperature. Temperature may also affect survival by influencing the rate of utilization of food reserves when food is present in limited quantities. This is particularly evident in blood-sucking insects such as the tsetse flies, Glossina spp. (Diptera, Muscidae), which depend on stored food reserves between meals. Increases in temperature shorten the survival period between meals. Obviously, if a fly use up its reserves from one meal before it is able to obtain another, it will perish.

Temperature exerts a strong influence on the reproduction and rate of development of insects. As with lethal limits of temperature, insects also have definite tolerable ranges of temperature in terms of reproduction and development beyond which neither will occur. Four example, the temperature range in which the eggs of the beetle Ptinus (Coleoptera, Ptinidae) will develop it between 5 and 28°C. Pediculus (Anoplura, Pediculidae) fails to loy eggs below 25°C. For a given species the range in which development will occur is probably somewhat broader than that in which reproduction will be successful. Within the tolerable ranges, egg development, oviposition rate, and the rate of larval and development usually increase with increasing temperature. For example, the duration of pupal life of the mealworm beetle, Tenebrio mollitor (Coleoptera, Tenebrionidae), is decreased 180 hours (from 320 to 140) as the temperature is increased by 12°C (from 21 to 33° C.)

The specific ranges of tolerance and influence of temperature on rates of reproduction and development vary among members of the same species. In addition, the span of ranges varies from species to species. Within each range of tolerance is an optimum zone in which the rates of reproduction and development are maximal. For example, the oviposition rate of Toxoptera graminum (Hemiptera-Homoptera, Aphididae) increases with increasing temperature to a maximum of approximately 25°C and then fall off. Most labouratory studies of the effects of temperature on reproduction and development have been carried out at constant temperature. The results of such studies do not necessarily reflect what would occur under uncontrolled field conditions. Periodically fluctuating temperature, which is characteristic of field conditions, tends to induce higher rates of development than would occur at constant temperature. For example, "grasshopper eggs kept at a variable temperature showed an average acceleration of 38.6 per cent and nymphs an acceleration of 12 per cent, over development at comparable constant temperature".

The distribution, horizontal and vertical, of an insect species is often greatly affected by temperature. In temperate zones the northern extreme of a given insect's distribution is commonly determined by low-temperature extremes. When northern limits are determined in this manner, there is usually a zone somewhat below the extreme limits in which the overwintering stage is killed but which is repopulated during the warm season. Proceeding southward from such an area, a greater and greater percentage of the overwintering individuals survive. For example, the corn earworm, Heliothis zea (Lepidoptera, Noctuidae), in eastern North America must become completely re-established during the warm season each year in Canada and to a progressively lesser extent in the United States proceeding southward. The ability to become cold-hardy also plays an important role in determining the northern most limits of distribution. The number of generations per year often varies between the northern and southern regions of an insect's range. For example, the corn earworm completes a generation approximately every 36 days as long as weather permits. Thus, in its northern range it may only complete two or three generation per year, whereas in the southern portion it may complete several more, breeding the year round. One would expect a gradation in the actual number of generations between the northern and southern extremes of the range. Temperature also affects, to a greater or lesser extent, the rate of dispersal of an insects species.

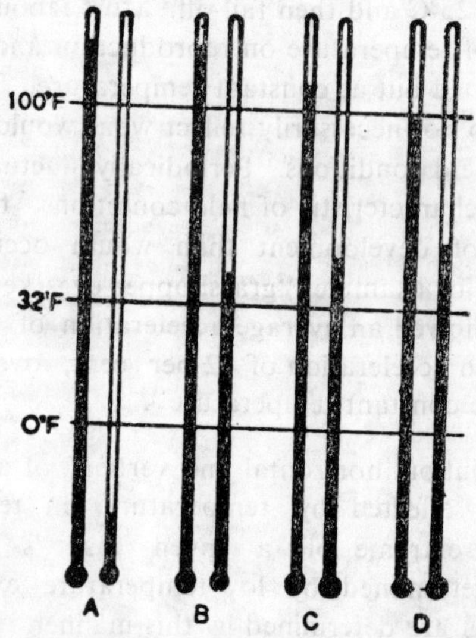

FIG. 2.4: Tolerable temperature ranges for various aspects of a hypothetical insects life. The left-hand thermometer of each pair registers the maximum tolerable ternperature: the right-hand one, the minimum tolerable temperature. A. Survival. B. Mobility. C. Development. D. Reproduction.

To this point we have been discussing the various ways temperature influences insects. We now want to consider the actual temperature of an insect and the factors that may influence it. We have established the idea that the temperature, or any of several other parameters, may vary substantially even within a small area (*e.g.*, different parts of a plant). Likewise, the temperature of an insect in one part of a habitat may not be anywere near the same as one in another and may not be the same as the ambient temperature. On the other hand, the temperature of an insect under controlled laboratory conditions usually reflects the ambient temperature.

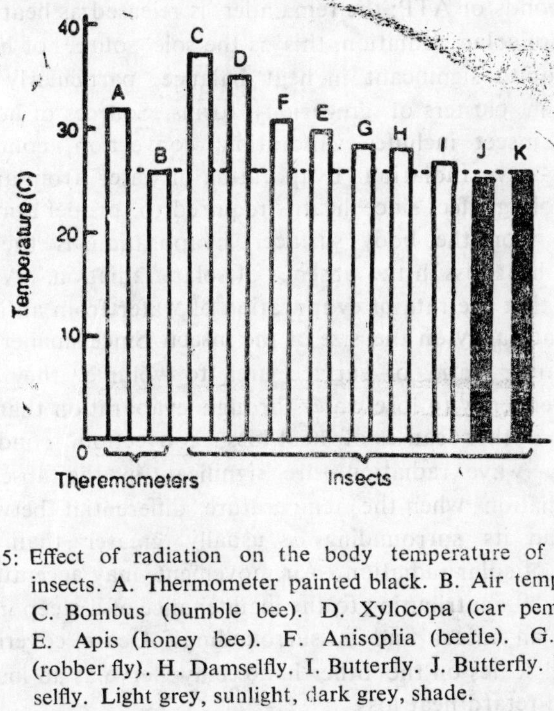

FIG. 2.5: Effect of radiation on the body temperature of selected insects. A. Thermometer painted black. B. Air temperature. C. Bombus (bumble bee). D. Xylocopa (car pemer bee). E. Apis (honey bee). F. Anisoplia (beetle). G. (Asilus) (robber fly). H. Damselfly. I. Butterfly. J. Butterfly. K. Damselfly. Light grey, sunlight, dark grey, shade.

The temperature of an insect depends on the sources of heat gain and loss operating under a given set of circumstances. The major sources of heat gain are solar radiation and metabolic heat. Solar radiation may cause the temperature of an exposed insect to be significantly different from the ambient temperature. The effect of solar radiation on the temperature of an insect is influenced by such factors as size, larger insects being more affected than smaller ones; colour, darker colours absorbing more radiation than lighter ones: shape, the more surface directly exposed to radiation, the greater the absorption; and orientation with respect to the sun, some insects orienting such that a larger or small amount of body surface is exposed.

Metabolic heat results from the breakdown of complex organic molecules. Part of this energy is stored in the high—energy bonds of ATP; the remainder is released as heat. In the absence of solar radiation, this is the sole source of heat and can be quite significant in heat balance, particularly during flight or in clusters of gregarious forms. Sources of heat loss from an insect include evaporation, convection, conduction, and long-wave radiation. Evaporation of water from an insect has a cooling effect since heat is required to propel a molecule of water from the body surface. Evaporation is the major cause of heat loss in the absence of solar radiation. As stated already, that the rate of evaporation of water from an insect is dependent partly on the size of the insect. Since smaller insects have a larger ratio of surface area to volume, they have a greater tendency to lose water through evaporation than larger ones. The other causes of heat loss—convection, conduction, and long—wave radiation—are significant in the absence of solar radiation when the temperature differential between an insect and its surroundings is usually greater than in the presence of solar radiation. Air movement may accentuate the heat loss by contributing to the maintenance of a steep gradient between an insect and its surroundings. Dense coverings of hairs and scales on the other hand, may serve as an insulating layer and retard heat loss.

Moisture

The water content of insects varies from less than 50 per cent to more than 90 per cent of total body weight. Variation occurs both between different species and between different life stages of the same species. Soft-bodied insects such as caterpillars tend to have comparatively large amounts of water in their tissues, whereas many insects with hard bodies (*i.e.*, relatively thick cuticles) tend to have somewhat lesser amounts. Active stages commonly have a higher water content than dormant stages. In most instances it is critical that water content be maintained within certain limits, which are influenced by several other environmental factors (*e.g.*, temperature, pressure, air movement, available surface water). If the limits of tolerance under a given set of circumstances are exceeded, an insect either perishes or many of its activities are seriously impaired.

Environmental moisture factors of major significance are precipitation, humidity, condensation, and available surface water. Rainfall is the most common and widespread form of fluid precipitation. Show is the most common form of solid precipitation; hail and sleet are less common forms. The annual and seasonal amounts of precipitation are primarily determined by the movements of large masses of air and by topographical characteristics. Thus, there are very "wet" and very "dry" regions and innumerable gradations between. Humidity refers to the amount of water vapour in the air and depends on temperature and atmospheric pressure. Condensation (dew, fog, and white frost) occurs when the atmosphere becomes saturated with water vapour. Saturation is the result of the relative humidity approaching 100 per cent or the temperature dropping below the dew point (the point at which the relative humidity becomes 100 per cent).

Available surface water is related to all the other moisture factors and to the nature of the substrate (soil, leaf surface, bark and so on). It is well known that soils vary on their water-holding capacity and the rapidity with which water runs off or soaks in. All these moisture factors influence the water balance of terrestrial insects. What, then, are the environmental factors that influence the water balance of aquatic insects? The "wetness" or "dryness" of an aquatic environment is a function of the osmotic pressure. Thus, insects living in fresh water must cope with a comparatively "wet" environment; those in brackish and salt water (*e.g.*, many of the salt-marsh forms, such as the mosquito Aedes sollicitans) are living in a very "dry" environment. Compounding the water-balance problem in insects is the fact that under other than carefully controlled laboratory conditions the moisture factors continually change. Obviously, an insect living in a given locale must be able to survive the extremes of these changes. As with temperature and the other weather parameters, one must think in terms of microenvironments. For example, the amount of rainfall measured at the edge of a forest hardly reflects the actual amount of water reaching an insect living in a tree hole or on the underside of a leaf. All the moisture factors vary both temporally and spatially. For example, relative humidity varies with location, time of day or year, topography, vegetation, and so on, and commonly tends to be comparatively high during the night and lower during the day. It may also be different at different heights above the ground.

As with temperature, there is an optimal moisture range in which a given species thrives. Mortality may occur under conditions of excessively low environmental moisture content for the active stages of most insects, and under conditions of excessively high moisture content for many. Death under very dry conditions is generally due to excessive water loss. Under conditions of excessive moisture the causes of death are more variable and may be direct or indirect. For example, drowning may be the result of excessive moisture, as is the case with overwintering pupae of the moth Heliothis zea during wet years. Survival is indirectly affected by very wet conditions that favour the spread of viral, fungal, and bacterial diseases and by any negative effects the excessive moisture might have on the food of a given insect species. The fall webworm, Hyphantria cunea, the armyworm, Pseudaletia unipuncta, and the gypsy moth, Porthetria dispar, succumb most readily to viruses when the weather is warm and the humidity is high. Whether high humidity affects the host or the development of the pathogen is not clear. However, many viruses do not develop rapidly unless temperatures of 21 to 29.4°C and relative humidities of 50 to 60 per cent are experienced.

If excessive moisture does not kill an insect, it may seriously affect the length of its life. For example, newly emerged adult migratory locusts, Locusta migratoria; live longer the lower the humidity. Since environmental moisture content may determine survival, it is commonly a major factor along with temperature in determining the geographic distribution of air insect species. For example, according to Rolston and McCoy (1966), the High Plains grasshopper reproduces and develops in an area of about 50,000 square miles, and from this center the adults may spread over an area perhaps twice as large. The region of endemic infestation lies in the short-grass belt of the High Plains where the average winter temperature generally falls between 28° and 38°F and the average annual precipitation ranges from about 15 to 18 inches. That the High Plains grasshopper flourishes in only a part of the short-grass belt indicates that climatic conditions, rather than host availability, limit the range of the insect.

Extremes of environmental moisture content directly influence many of the activities of insects, including feeding, reproduction, and development. Spruce budworm larvae stop feeding when the air becomes saturated with water and the tsetse, Glossina tachinoides, does not feed on its vertebrate hosts when the relative humidity is about 88 per cent. Newly emerged adult migratory locusts do not produce eggs below about 40 per cent relative humidity. Generally low humidities adversely affect the rates of oviposition, which increase as humidity increases. The rate of development may be decreased by extremes of moisture content or development may be halted altogether. Under very moist conditions silkworm larvae fail to pupate. According to Bursell (1974) the incubation time for eggs of the spider beetle Ptinus, under a constant temperature of 20°C, is 15 days at 30 per cent relative humidity and 10 days at 90 per cent. He further points out that generally higher humidities are more favourable for embryonic stages than low humidities. Two sets of hypothetical curves summarize the points made in this and the preceding paragraphs.

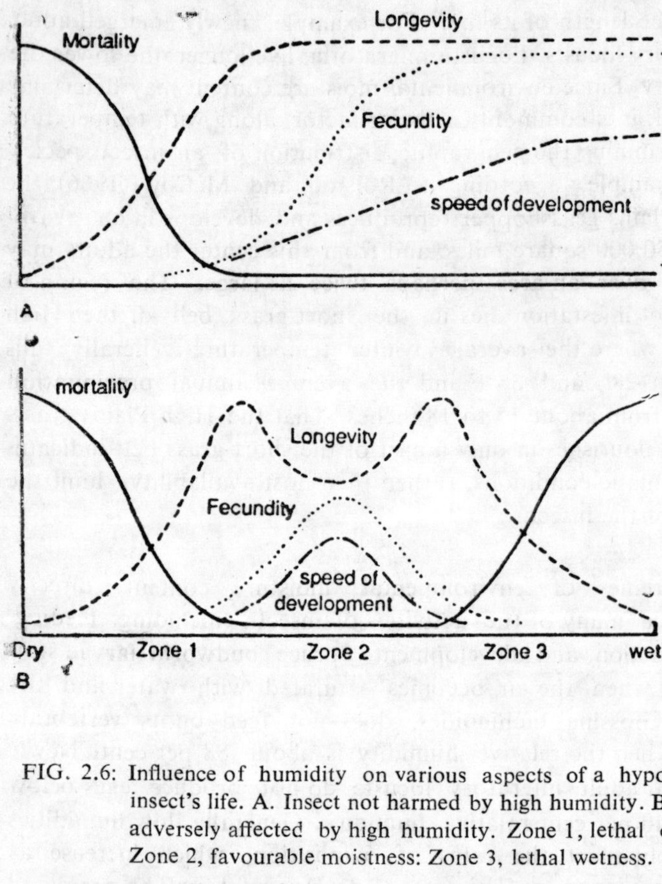

FIG. 2.6: Influence of humidity on various aspects of a hypothetical insect's life. A. Insect not harmed by high humidity. B. Insect adversely affected by high humidity. Zone 1, lethal dryness: Zone 2, favourable moistness: Zone 3, lethal wetness.

The effects of environmental moisture content are often strongly modified by other weather factors. Temperature and moisture, in particular, interact to a large extent in their effects. Thus, temperature exerts a relatively great effect on insects at the extremes of moisture conditions and *vice versa*. For example, the boll weevil is more tolerant of higher temperatures at comparatively low humidities than at high humidities. Another good example of the interactions of temperature and moisture is their combined effect on the rate of development of insects. As mentioned earlier, air movement, atmospheric pressure, availability of surface water, and other factors may also influence the effects of moisture conditions on insects.

Insects are adapted in a number of ways and to varying degrees to cope with changes in environmental moisture conditions. Dormant stages are often well suited for exposure to drought. The eggs of many species of mosquitoes (Aedes, Psorophora, and others) can withstand prolonged drying in air. Many insects become dormant in response to different environmental conditions and in this state are able to withstand long periods of drought. For example, the potato beetle, Leptinotarsa decemlineata, becomes quiescent in response to dryness and can survive in this resting, desiccated state for months. Such a response ensures that when eggs are laid, the larvae will be exposed to environmental conditions more favourable for survival. Hinton (1951, 1960) describes a chironomid (Diptera, Chironomidae) larva, Polypedilum vanderplanki, that is able to survive almost complete dehydration for several years. These insects show no visible signs of metabolic activity and are described as being in a state of cryptobiosis.

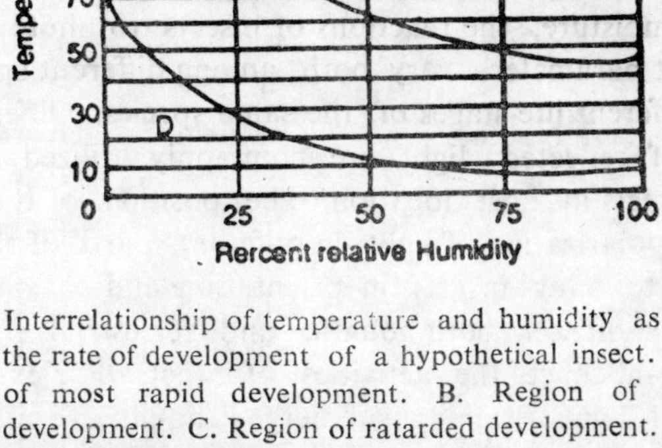

FIG. 2.7: Interrelationship of temperature and humidity as they affect the rate of development of a hypothetical insect. A. Region of most rapid development. B. Region of favourable development. C. Region of ratarded development. D. Region of no development.

Most insects offset the influence of dryness and other factors promoting water loss by drinking water or taking it in with food. Time is important in the survival of active insect under adverse moisture conditions. Many can survive extensive desiccation for extended periods. Some insects that live under extremely dry conditions, such as mealworm beetle larvae (Tenebrio molitor, Tenebr onidae, Coleoptera), are able to utilize metabolic water (*i.e.*, water that becomes available as a result of the metabolic breakdown of food materials). Other insects, such as some Thysanura (silverfish and relatives), are able to absorb moisture directly from the atmosphere.

Light

Light is of direct important to insects more as an environment point of reference than as a survival factor since its parameters (photoperiod, illuminance, wavelength, etc.) are more or less constant and it is seldom, if ever, directly lethal under natural conditions. In the non-equatorial region of the earth there is a regular change in photoperiod due to the tilt of the earth's axis of rotation 23½° from vertical to the imaginary plane that passes through the sun and the earth's orbit. Regular change in photoperiod serves as an annual clock for insects and is used by many to maintain synchrony with the seasons and their host plants. Photoperiod is one of the major stimuli that induces diapause. The daily cycle of dark and light with crepuscular (dawn and dusk) periods in between also serves as a clock by which the feeding, mating, etc., are regulated. As with temperature and moisture, the reactions of insects to photoperiod and other light parameters vary both among different species and among different life stages of the same species. Different wavelengths of reflected light are commonly utilized by plant-feeding insects in host location. The position of the sun and degree of polarization of light in different parts of the sky are important to many insects in orientation and navigation. The effect of light on both aquatic and terrestrial plants may indirectly influence the activities of insects. For example, the amount of light reaching submerged aquatic vegetation will affect oxygen—generating photosynthesis and in turn affect the oxygen concentration in the water, which then influences aquatic insects.

Other Factors

Other environmental factors that under some circumstances influence insects include currents in air and water, gases dissolved in water, air composition, electricity and ionizing radiation. Air and water currents are determined to a large extent by physiographic conditions. Air movement is modified by trees and other vegetation and by anything else that may block or redirect it. Currents in water are influenced by such factors as water volume, slope of stream bed, and temperature differences between the surface and various depths. Wind is a very effective agent in the distribution of insects—aphids, leafhoppers, and others being blown for hundreds or even thousands of miles. Air movement may be directly responsible for the death of insects in two ways. First, severe wind and heavy rain together may cause mortality. Second, movement of air above a surface where evaporation is occurring (*e.g.*, insect cuticle) increases the gradient of water vapour concentration and hence tends to increase the rate of evaporation. Other factors being constant, the rate of evaporation is proportional to air movement. On the other hand, air movement may be beneficial if humidity is high.

Water currents often determine which species of insects will live in a given area. For example, the various genera of mayflies may be classified into still-or rapid-water forms. The legs and bodies of these insects are appropriately adapted (*e.g.*, legs capable of clinging and stream—lined bodies associated with fast—moving water). Black fly larvae fasten themselves to stones or other stationary material in the water. Several caddisflies (Trichoptera) attach their cases to submerged objects. Many aquatic insects (*e.g.*, mosquito larvae) are unable to survive in moving water. Another important aspect of currents in the aquatic environment involves the circulation of dissolved gases, salts, and nutrients. The insect fauna in a given location are often determined to a large extent by the amount of dissolved oxygen. For instance, caddisfly and mayfly larvae may be found under conditions of relatively high oxygen concentration, midge and black fly larvae (Diptera; Chironomidae and Simuliidae, respectively) at somewhat lower concentrations, and certain mosquito (Diptera, Culicidae) and other fly larvae at very low concentrations. Dissolved oxygen is determined by water movement, wind, water splashing, photosynthesis in aquatic plants, and so on. Other dissolved gases that may be of importance are carbon dioxide (a highly soluble waste product of cellular respiration) and nitrogen.

The gaseous composition of air is remarkably constant and is probably not a significant limiting factor for terrestrial insects and other animals. However, the gas content in cavities in flowers may be quite different from the surrounding air, and a large number of insects live in such cavities. Further, the composition of air in the middle of a thick canopy of vegetation is not necessarily the same as in open air, nor is it necessarily the same throughout the day since temperature and the photosynthetic activities of plants vary during the course of a day. For example, measurement of the carbon dioxide concentration above fields of wheat and clover at different times of day revealed that carbon dioxide decreased toward the middle of a day and reached a maximum between midnight and 6 A.M. Electrical factors have seldom been taken into account, although they may have a direct effect on insects. Under natural conditions, the ionization of air and atmospheric potential may vary and may affect certain activities of insects. For example, the ionization of air has been found to modify the flight activity of a blow fly, Calliphora sp. An increase in the number of ions causes a temporary increase in activity; for example, flight in Drosophila is abruptly reduced for a short period of time by sudden exposures to a potential gradient of 10-62 volts cm. Under natural conditions the atmospheric potential falls rapidly between the ground and a few meters above the ground.

Electromagnetic radiation bombards the earth continuously. Radiation in the visible and near visible region plays many important roles in life processes; for example, it affects photosynthesis and is involved in vision. Radiation of shorter, more energetic wavelength also influences living organisms. They highly energetic radiation arises from cosmic, solar, and terrestrial sources. Most radiation from extraterrestrial sources is absorbed or reflected by the atmosphere and deflected by the magnetic fields, but a significant amount gets through none the less. Cosmic radiation probably varies considerably from time to time as a result function of changes in the earth's magnetic fields, the occurrence of nearby supernovae (exploding stars), and so on. Such changes in the intensity of cosmic radiation may have played (and may continue to play) a major role in biological evolution by inducing drastic increases in mutation rates at times during the earth's history. Most terrestrial sources of highly energetic radiation occur in trace amounts throughout the earth's crust. However, since the beginning of this century. man has learned how to release immense quantities of radiation by "splitting the atom." The inherent dangers of this activity are well known Entomological interest in high—energy radiation stems from:

1. The usefulness of radiation as a basic research tool, for example, radioisotopes are used as tracers in the study of bio-chemistry, and radiation is used to induce physiological, morphological, and genetic changes with the hope of further understanding biological processes.
2. The usefulness of insects as model biological systems in evaluating the harmful effects of radiation.
3. The applied value of high-energy radiation in inducing sterilization or deleterious genetic changes in insects pests. For information and literature sources regarding the influence of high-energy radiation on insects.

Food

This section deals with the ways insects may be influenced by variations in quantity and quality of food resources. As a group, insects are essntially omnivorous. As individual species, however, they exhibit mueh variation, some being extremely selective and perhaps relying on only one or a few kinds of food (including host plants and animals and prey) and others taking advantage of many kinds of food. Whatever the habits of a given species may be, abundance and quality of food may play an important role in survival, longevity, distribution reproduction, speed of development, and so on.

Quantity of Food

Animal populations that consume all or most of their food resources are few compared with those that do not. Thus "absolute" shortage of food is probably not an important limiting factor for most populations. However, such shortages may occur in patches throughout the distribution of a given population. In these circumstances there is "effective" shortage of food.

There are many possible causes of "absolute" and "effective" shortages of food. Among them are large numbers of individuals per unit quantity of food (interaspecific competition), more than one species consuming the same food material(s) (interspecific competition), species that influence the food of other species without consuming it, and other environmental factors. The first instance might occur, for example, among insects during the actively growing larval stages. When there is a relatively small number of individuals per unit quantity of food, there is enough for all to grow to adulthood. However, as numbers increase, a point is reached where there is insufficient food for all individuals present to grow to maturity; only those that develop most rapidly reach adulthood, the slower developers do not. Obviously, two or more species consuming the same food will reduce the total amount available for any one of them. Species that do not actually consume the food supply of another may exert a positive or negative influence on that supply. For example, many of the sanitary practices of humans, such as incineration of garbage and treatment of waste materials, destroy the food supply of many insect species, especially cockroaches and flies. On the other hand, failure to carry out these procedures increases the same food supply, enhancing the survival of insects that thrive in such materials. Microbes that are pathogens of an insect that is the prey of another may reduce the population size of the prey species sufficiently to influence the predator species, an epizootic of the fungus Entomophthora so reduced the numbers of the caterpillar Plutella that the predators (Angitis spp. and others) suffered severely from a shortage of food where it had been abundant before the outbreak of disease. A similar relationship may exist between a plant species, the insects that feed on it, and microorganisms that are pathogenic to members of the plant species. In fact, any environmental factor, biotic or abiotic, that has the effect of reducing populations of particular plants or animals will probably cause the reduction not only of animal populations that utilize these plants or animals for food but also of the parasites and predators of these animals. Monophagous insects are more likely to be affected by such reductions in food supply that polyphagous species.

Examples of causes of "effective" shortages of food include accidental separation from the food source, certain behavioural traits, and effects of feeding activities on animal hosts. It is not difficult to imagine an ectoprasite starving to death when separated from its host. Tsetse flies (Diptera, Muscidae) remain close to tall bushes and will feed on vertebrates that are in the vicinity of these bushes, but will not feed on the same or other vertebrates in open grassland. Where a single or a few blood—feeding insects might successfully feed to satiation on a host, larger numbers may elicit an avoidance or destructive response on the part of the host, resulting in very few, if any, successfully feeding. Thus, although a host may have ample blood for several hundred tsetse flies, mosquitoes, and so on, it may tolerate the feeding activities of only a small fraction of that number. An absolute shortage of food could result from the food plant or animal being completely destroyed within a localized region, but it is unlikely that total destruction of a food plant or animal could occur in this way. In response to a food shortage, individuals may survive but be much smaller than they would have been with more food. However, as numbers of individuals increase, a point is eventually reached where fewer and fewer individuals are able to survive, and finally none survive to adulthood. Another frequently encountered result of food shortage is cannibalistic behaviour. One well-known species in which cannibalism is known to occur is the confused flour beetle, Tribolium confusum (Coleoptera, Tenebrionidae). Cannibalism may also occur when there is overcrowding, accidental injury to an individual, or during the vulnerable period immediately following ecdysis but before the cuticle has become sclerotized (*i.e.*, the teneral state). Several adaptations that promote survival under conditions of food shortage have evolved, particularly in response to regularly occurring shortages. Probably one of the most important is migration. Insects that respond to food shortages by migrating are much more likely to discover a fresh food source than those that do not. Many adult insects, fruit flies and house flies, for example, lay their eggs in situations where the food supply is in excess relative to the numbers of individuals feeding on it. The beahviour of the adults in egg deposition essentially guarantees abundant food for their offspring. Polyphagy on the part of a parasite, predator, or herbivorous insect also helps ensure against the likelihood of being exposed to dangerous food shortages.

Quality of Food

Nutritional requirements of insects in general are discussed in Chapter 4. Egg production and larval development are especially susceptible to variations in the nutritive quality of food. Survival, longevity, and size may also be strongly influenced by the quality of available food. Many insect species store sufficient nutrients during the larval stage to accomplish adult activities (copulation, egg production, and oviposition). The adult lives of these species are of comparatively short duration, and they commonly do not feed at all. For example, mayflies (Ephemeroptera), which live only long enough to copulate and deposit their eggs rely entirely on larval reserves. In other species, larvae may store nutrients sufficient for egg production, but the adults must ingest water and carbohydrates to survive.

Many mosquitoes and other true flies (Diptera) fall into a third category in which adults usually need to ingest a complete diet (especially protein) in order to sustain life and produce eggs. For mosquitoes and other blood-feeding forms, blood meals afford the necessary nutrients. The queens of the social Hymenoptera and termites also fall into this category. These insects lay eggs for almost all of their adult lives and require food constantly. Commonly, this food is provided by a worker caste, which in turn may obtain food by foraging.

An example of the effect of quality of food on the rate of development is given by Folsom and Wardle (1934). Both bananas and horse manure are suitable foods for house fly maggots. However, at approximately 21°C, the maggots complete their development on horse manure one to nearly two weeks sooner than on bananas. The quality of ingested food in some instances influences the outcome of development; for example, the differences between workers and queen honey bees depend entirely on the diet that each receives during larval development. Both receive the same food for the first two days of larval life—royal jelly, a substance secreted by glands that open into the mouths of workers. After the third day, larvae destined to become workers receive a diet containing increasing amounts of honey, whereas the queen larvae continue to be fed royal jelly. Another example of food quality affecting development is found in aphids (Hemiptera—Homoptera, Aphididae). Changes in the quality of food may induce the appearance of winged form and migratory behaviour.

A PLACE IN WHICH TO LIVE

A place where an organism lives is its habitat. This "place" may be an entire field or forest, the shoreline of a stream, the undersides of leaves of a particular kind of plant, among others. A given species is typically found living in essentially the same kind of surroundings throughout its distribution, although there may be variation within the limits of tolerance of the species. The degree of suitability of a given habitat will vary with the extent to which it satisfies the needs within these limits of tolerance. Each species of the community of organisms living in a given habitat requires certain resources (*e.g.*, food, shelter, breeding sites, favourable temperature, favourable humidity space, and time) in sufficient amounts and of sufficient quality to survive and reproduce. That is, every species occupies a particular physical location at a given time and does a given thing (eats, copulates, rests, etc.) at that place and time. A compliation of all needed resources, both in qualitative and quantitative terms, describes the ecological niche of a species. A community is composed of individuals of different species segregated into different niches.

Insects commonly occupy a wide variety of niches in almost every kind of terrestrial and freshwater habitat. About the only habitats not broadly colonized by insects are those within the ocean; although many insects from different others are found in intertidal marine habitants, and some are found on the open ocean (*e.g.*, several species of water striders, Halobates, Hemiptera, Gerridae. However, there are no known marine insects that spend their entire life cycle submerged. Janzen (1977) provides an interesting discussion of the possible reasons why insects in particular have been able to occupy so many different niches.

OTHER ORGANISMS

Other organisms include individuals of the same or different species as part of each other's environment. The topic thus divides into intra—and interspecific interactions.

Intraspecific Interactions

The ways in which members of the same species interact are related in part to population density. There seem to be advantages associated both with relatively low and relatively high densities. Examples of circumstances in which high population density may be advantageous include mate finding and survival of potential predation. In a situation where the probability of finding a mate is decreased owing to low population density, the result would be fewer fertilizations and hence a lower average fertility. Whether a given population density is advantageous or disadvantageous is relative and depends upon other environmental influences. This is particularly apparent with regard to such available resources as food and shelter. The maximum population of a given species an area can support depends on the abundance of these resources in that area. When resources are limited, intraspecific competition may become increasingly pronounced as the maximum population is approached and may result in the death or emigration of individuals. Other environmental influences may, however, serve to keep a population well below a level where competition for limited resources would occur.

Interspecific Interactions

Interspecific interactions may be conveniently divided into competition, symbiosis, predator-prey interactions, herbivore-plant interactions, and indirect interactions. Interspecific competition comes about when the needs of two or more different species for a given resource (food, shelter, etc.) coincide—when their niches overlap. In situations where two species with essentially identical life needs are brought together, one species may be expected to have a competitive advantage and eventually to eliminate the other. Stated another way: "no two species can occupy the same niche at the same time for very long." The segregation of the different species in an ecosystem into different niches is explained on the basis of the operation of competitive exclusion. A good example of competitive exclusion is the outcome when two species of flour beetles (Tribolium castaneum and T. confusum) are placed in the same container of flour. If the container is maintained under conditions of high humidity and temperature, T. castaneum invariably wins out. However, maintaining the beetles under conditions of low humidity and temperature results in a "victory" for T. confusum. In the absence of the other, either species can be maintained indefinitely under wet and warm or dry and cool conditions.

Symbiosis is used here to mean a close association between two different species. This association may be mutually advantageous (mutualism), disadvantageous for one of the species (parasitism), or advantageous for one without harm to the other commensalism).

There are many examples of mutualistic associations between insects; one is seen when ants actively care for and protect aphids, which in turn excrete honeydew, that is ingested by the ants. The relationships between fungus-growing ants and fungi are good examples of mutualism between insects and other kinds of organisms. Additional examples include the relationships between microbes and insects relative to digestion and nutrition. A recent review deals specifically with the intracellular symbiotes of the Hemiptera-Homoptera. Koch (1967) discusses endosymbionts of insects. There are numerous examples of mutualistic relationships between plants and insects where the plants provide food for an insect that in turn pollinates the plants. Among the organisms that parasitize insects are insects themselves micro-organisms, mites, and nematodes. In some instances microbes and nematodes parasitize both insects and vertebrates. In these cases the insects may act as vectors, carrying parasites to their vertebrate hosts or plants hosts.

Well over 1000 microbes, most of them pathogens (disease-causers), have been described in insects, and new ones are being added regularly. According to N.S.A. (1969), the current list include "90 species and varieties of bacteria, 260 species of viruses and rickettsiae, 460 species of fungi, 255 species of protozoa." These microbes may gain entrance into insects orally or via wounds in the integument, or they may be capable of actively penetrating the integument (particularly fungi). In some instances pathogens that have gained entrance mey be passed transovarially (via the egg) to offspring, as are certain arboviruses in mosquitoes. Interest in pathogenic microbes associated with insects has largely been focused on their potential use in biological control. During the past 30 years or so, insect microbiology has grown rapidly as an entomological specialty.

Among the bacteria that infect insects, the best known is Bacillus thuringiensis, a spore-former that has been used with some success in biological control. It is pathogenic for nearly 200 species of pest insects, particularly Lepidoptera. Another member of the same genus, Bacillus larvae, causes American foulbrood, a serious disease of honey bee larvae. In addition to members of the genus Bacillus, certain species of Clostridium and Streptococcus are also insect pathogens. Many bacterial diseases of plants are mechanically transmitted by insects. Certain bacterial pathogens of vertebrates are also trasmsitted by insects, for example, those that cause plague (Yersinia pestis) in human.

Viruses are obligate cellular parasites composed in part of DNA or RNA. Those associated with insects are generally not as well known or understood as the bacteria. They are classified on the basis of the part of the host cell in which they develop, the presence or absence and morphology of an inclusion body, and the kind of nucleic acid (DNA or RNA) present. Some insect viruses produce granular inclusion bodies; most produce crystalline poloyhedral bodies in the nucleus or cytoplasm of an infected cell, Virus particles may be alone in an inclusion body or in packets. Several viruses that are pathogenic for insects have been identified and described. Most have been found among Lepidoptera, although several also are known from the Hymenoptera, Diptera, Coleoptera, and Neuroptera. It is interesting that a virus may act in conjunction with a bacterium to produce a disease syndrome not produced by the virus of bacterium separately. Such is the case in two silkworm diseases, gattine and flacherie, Streptococcus bombycis and

Bacillus bombycis, respectively, being the bacteria involved. Insects, especially aphids and leafhoppers (Hemiptera-Homoptera), are important vectors of viruses pathogenic for plants. Insects also serve as vectors of several viruses that attack humans and other vertebrates, notably the arboviruses (arthropod-borne viruses), including those that cause yellow fever, dengue, and encepthalitis. Microbes somewhat similar to bacteria, but lacking cell walls, mycoplasmas, have recently been found to cause the "yellows diseases" of plants, such as aster yellows. Leafhoppers (Hemiptera-Homoptera, Cicadellidae) are involved in the transmission of these organisms.

Some rickettsiae are insect pathogens. In size they lie somewhere between viruses and bacteria, but, unlike viruses, they are capable of independent metabolism in vitro. Those rickettsiae that are pathogenic for insects are very slow in killing their hosts. They have been grown in mammalian-tissue culture and have killed white mice upon injection. Owing to slow kill of insects and potential danger for mammals, they are unlikely ever to be used in insect control. Rickettsia prowazekii (causes epidemic typhus) and R. prowazekii mooseri (causes murine and endemic typhus) which are pathogenic for humans, are vectored by body lice and fleas, respectively. There are more species of fungi that are known to be pathogenic to insects than of any other group of microbes. These entomogenous fungi have been described from several groups of fungi. Species in the genus Beauveria have been used in attempts at biological control. However, the success of fungus infections in insects (*i.e.*, whether they become established and kill an insect) depends to a great extent on weather conditions, and hence these microbes would likely be undependable as agents of biological control. Although insects are apparently not vectors of fungi that cause human or other vertebrate diseases, they do serve as mechanical vectors of fungi pathogenic to plants. An example is the fungus that causes Dutch elm disease, which is vectored mainly by the European elm bark beetle, Scolytus multistriatus. Madelin (1966) discusses fungi that are parasites of insects.

A large number of protozoa, especially Microsporidians are known to kill insects but, like rickettsiae and fungi, they act very slowly. No doubt many of those slow-acting microbes weaken their hosts and make them more susceptible to the effects of other environmental components. Insect pathogens have been described in the protozoan phyla Sarcodina (amebas), Mastigophora (flagellates), and Sporozoa (sporozoans). The microsporiodian Nosema bombycis (Sporozoa) causes the "pebrine" disease of silkworms. The elucidation of this relationship by Louis Pasteur in the nineteenth century stands as a classic in microbiological research. Other well-known microsporidian insect pathogens include Nosema apis and N. pyraustai, which attack the honey bee and European corn borer, respectively. A mosquito, Culex erraticus, parasitized by the microsporidian Thelahania minuta, is shown in Figure 9-10. Protozoans are common partners in mutualistic relationships with insects; we have already discussed those found in the gut of termites, which provide cellulase and enable their hots to digest cellulose. Several protozoa pathogenic for humans and other vertebrates are vectored by insects. Examples include the malarias (Plasmodium spp.) vectored by Anopheles spp. mosquitoes (different forms of malaria occur in birds, monkeys, and humans), American trypanosomiasis (Chagas' disease) caused by Trypanosoma cruzi and vectored by blood-sucking bugs (Hemiptera, Reduviidae), and the two forms of African sleeping sickness caused by T. rhodesiense and T. gambiae and vectored by tsetse flies (Diptera, Muscidae, Glossina spp.). The serious disease of cattle, nagana, which has retarded the development of major parts of Africa, is caused by the protozoan T. brucei, also vectored by tsetse flies.

Mites (Arachnida, Acarina) are very small arthropods, many species of which parasitize insects. For example, the larvae of members of the family Erythraeidae are parasites of insects and other arthropods and have been found attached to wing veins and other locations on the insect body. Acarapis woodi (Scutacaridae) causes the acarine disease of bees. This species either lives in the tracheae or on the external parts of the bee's body and weakens the host to the point where it can no longer fly and eventually succumbs.

About 100 species of nematodes (phylum Aschelminthes, round-worms); have been described as being associated with approximately 1500 insect species. In most instances, these "entomophilic" nematodes damage and eventually kill their host. Nematodes show some promise in the biological control of insects. A number of species that attack humans, for example, filarial worms, (Wuchereria bancrofti and others), and other vertebrates, for example, dog heartworm, spend part of their life cycles in insects (mosquitoes for the examples here), which serve as their vectors. Many relationship between two or more insect species or between insects and other organisms are commensal. For example, probably many of the microbes associated with insects neither harm nor particularly help their host but gain food and a place to live. Phoretic relationships are good examples of commensalism. In a phoretic relationship an individual of one species attaches to an individual of another and gains a mode of transportation. Insects may be transporters or riders, or both. Some chewing lice (Mallophaga) attach themselves by their mouth-parts to louse flies (Diptera, Hippoboscidae), some of which are also parasites of birds, and in this way move from one host to another. The torsalo fly, Dermatobia hominis, deposits eggs on other flies, such as mosquitoes, black flies, and house flies. When these insects come into contact with a potential vertebrate host (including man), larvae emerge from the eggs and penetrate the skin. Clausen (1976) discusses phoresy among entomophagous ("insect-eating") insects.

Certain insects, for example, gall midges (Diptera, Cecidomyiidae) and gall wasps (Hymenoptera, Cynipidae), include plants to form characteristic growths in which the larvae of these insects feed, grow, and develop. These growths are called galls and are made up entirely of plant materials. Depending upon the effect of galls on a plant, such a relationship could be viewed as parasitic or commensalistic. Certain fungi nematodes, mites, and other organisms also cause galls to develop, but insects are the most common gall inducers and the most elaborate and specific galls are those induced to form by insects.

Predators of insects include insects themselves: other Arthropoda, Such as mites, spiders, scorpions, and pseudo;-corpions; and vertebrates, including birds, reptiles, amphibian, fish, and mammals. Sweetman (1936) discusses predatory invertebrates in some detail. In addition to animals, a few plant species, such as thee Venus flytrap, pitcher plant, and the aquatic Utricularia vulgaris, trap, digest, and assimilate insects. Heslop-Harrison (1978) and Swartz (1974) discuss carnivorous plants.

A majority of insects feed on plants (*i.e.*, are herbivores). In most cases the association goes much further than simple herbivory, the plants providing shelter, oviposition sites, and so on. Insects may be influenced by other organisms indirectly. No doubt the greatest effects are those produced by humans purposely against pest populations and inadvertently against the vast majority of insects that are "innocent". The effects of modern agriculture, urban sprawl, and superhighways on various habitats are obvious, and the far-reaching and unfortunate effects of highly residual pesticides have become well known. The effects of air and water pollution on the environment are beginning to get the kind of attention they deserve. Insects themselves are sometimes polluters: "Predatory anthocorid bugs enter open galls and feed on a few of the aphids therein. The aphids which are not attacked die from the fumigant effect of chemicals produced by the stink glands of the bugs".

Adaptations Associated with Interspecific Interactions

The "goal" of every organism is to obtain enough food (matter and energy) to maintain itself long enough to reproduce, passing its genes to the next generation. As an organism proceeds from conception to reproductive maturity, it must constantly obtain food and at the same time avoid becoming

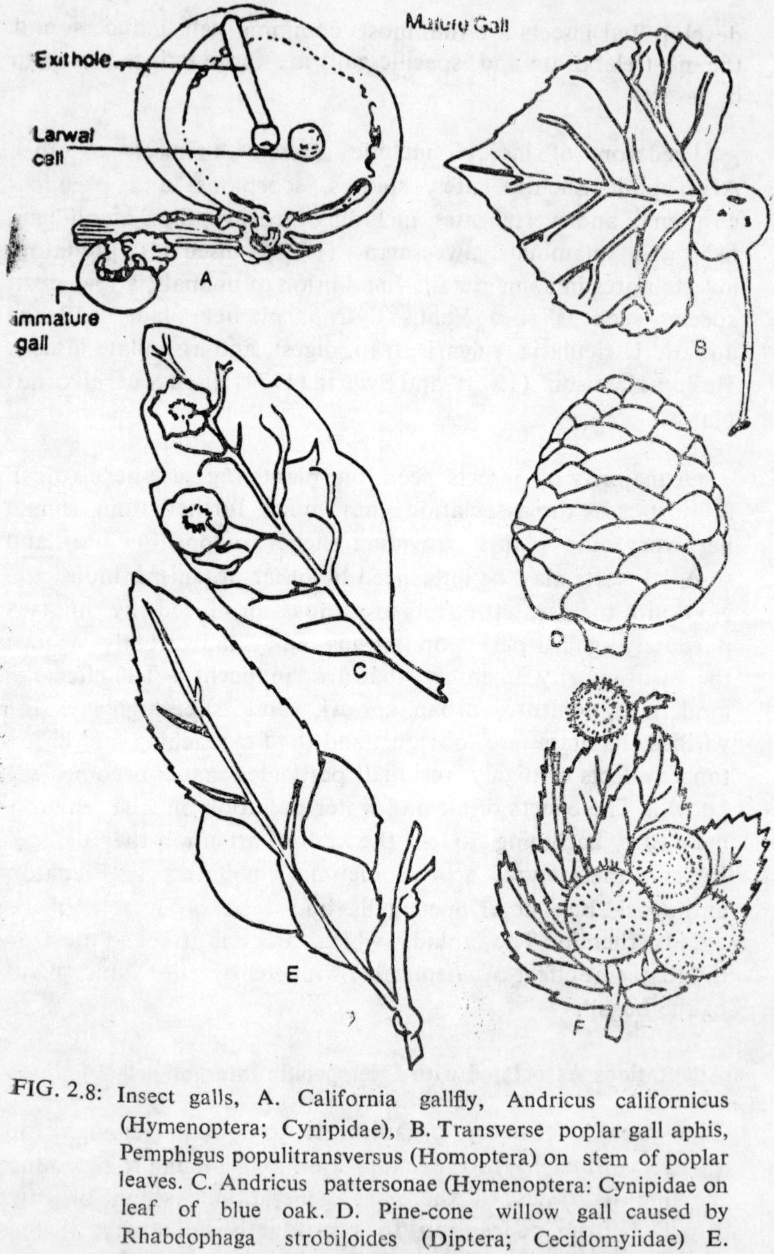

FIG. 2.8: Insect galls, A. California gallfly, Andricus californicus (Hymenoptera; Cynipidae), B. Transverse poplar gall aphis, Pemphigus populitransversus (Homoptera) on stem of poplar leaves. C. Andricus pattersonae (Hymenoptera: Cynipidae on leaf of blue oak. D. Pine-cone willow gall caused by Rhabdophaga strobiloides (Diptera; Cecidomyiidae) E. Willow apple gall caused by the swafly Pontania pomum (Hymenoptera; Tenthredinidae). F. Spiny rose gall caused by Rhodites puslulatoides (Hymenoptera; Cynipidae).

food for some other organism. Organisms that are unable to do this successfully, for whatever reason, are maladapted and are, in the long term selected against. Although a number of environmental factors are no doubt involved in the selection process, examination of specific adaptations often leads to the identification of specific selecting agents. Such is the case when we examine adapted one associated with many inter-specific interactions. Organisms that are closely associated ecologically often act as agents of selection reciprocally; that is, they act as selecting agents on one another. There is good evidence that many organisms have done so for long periods of geologic time. These organisms are said to have coevolved or to be coadapted.

Examination of several examples where coevolution has played a role reveals at least two types of coevolutionary interactions. Two (or more) different species may have co-evolved to the extent that they benefit one another in the quest for reproductive success; their relationship thus is mutualistic. In this case the partners have become (or are becoming) more and more successful in their interaction leading to sustained or improved reproductive success for both. Alternatively, the participants may have engaged and continue to engage in a kind of genetic warfare in which one member "evolves" a mechanism of offense providing increased access to the other as food, only to have the other "respond", in the evolutionary sense, with a defense, and so on through time. Such is the case in interactions between parasites (and parasitoides) and hosts, predators and prey, and herbivores and plants. An excellent example of coevolution that has resulted in a mutualistic relationship is that which exists between the bull's-horn acacia, *Acacia* cornigera, and the ant Pseudomyrmex ferruginea, Janzen (1967) describes this relationship as follows:

> The bull's-horn acacia . . . is a representative swollen-thorn acacia with well-developed foliar nectaries, enlarged stipular thorns, and small nutritive organs (Beltian bodies) borne at the tip of each leaf segment. The colony of P. ferruginea living in the enlarged stipules obtains sugars from the

foliar nectaries, and oils and proteins by eating the Beltian bodies. The workers patrol and clean the surfaces of the acacia, and bite and sting animals of all sizes that contact the plant. The workers maul any other species of plant that contact the acacia and in many cases, any that grow under the acacia. The colony attains a very size and up to 25 per cent of the workers may be active on the surface of the acacia both day and night. The larger the colony becomes, the smaller is the damage sustained by the plant from defoliators. The colony enhances its own probability of survival by protecting the acacia on which it is completely dependent for food and domatia.

Pollinating insects and the plants they pollinate also provide excellent examples of adaptations associated with long co-evolutionary relationships. Sexual reproduction in flowering plants is accomplished by the transfer of pollen from the anther of a male flower to the stigma of a female flower. When a pollen grain contacts a stigma, the male germ cell in the pollen grain unites with the female germ cell or egg, and a fertile seed develops. When the seed is exposed to the proper conditions in or on an appropriate substrate, it will give rise to a new plant. The transfer of pollen from a male to a female flower is accomplished primarily by the wind or by the activities of insects that associate with plants. Examples of wind-pollinated plants include cereal plants such as wheat and corn and many species of trees. The flowers of these plants are generally small with weakly developed petals, do not produce nectar, and produce dry pollen grains, which are easily picked up by the wind.

The relationships between plants and their insect pollinators are mutualistic in that the plants provide sugar-rich nectar and/or pollen for insects, which in turn serve as accessory plant "sex organs" in transmitting gametes from the male to the female plant (or flower). Recently, nectar of certain plants has been found to contain nutritive substances, such as certain amino acids and lipids, in addition to sugar. Insect-pollinated plants have evolved traits that make them attractive to effective

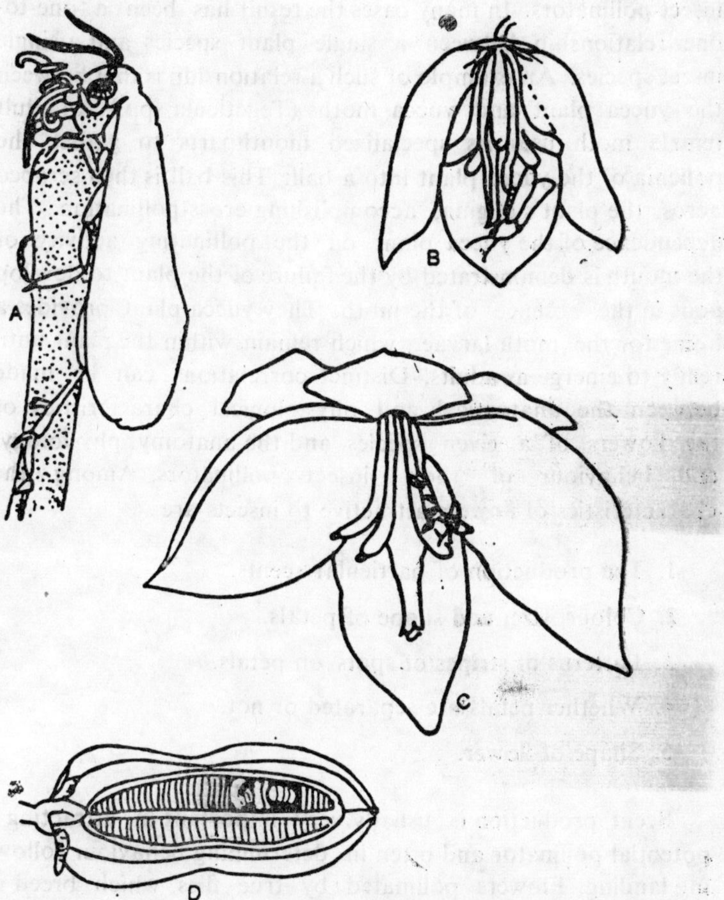

FIG. 2.9: Yucca moth and yucca plant. A. Female moth gathering pollen. B. Female scraping the pollen ball across the plant's stigma. C. Female ovipositing (several petals removed). D. Larva within mature pod (portion of pod removed). E. Pod with exit holes and constrictions caused by larvae.

insect-pollinators. In many cases the result has been a one-to-one relationship between a single plant species and a single insect species. An example of such a relationship is that between the yucca plant and yucca moths (Tegeticula spp.) An adult female moth uses its specialized mouthparts to gather the pollenia of the yucca plant into a ball. This ball is then scraped across the plant's stigma, accomplishing cross-pollination. The dependence of the yucca plant on the pollinating activity of the mouth is demonstrated by the failure of the plant to develop pods in the absence of the moth. They yucca plant provides a home for the moth larvae, which remain within the plant until ready to emerge as adults. Distinct correlations can be made between the anatomical and physiological characteristics of the flowers of a given species and the anatomy, physiology, and behaviour of their insect pollinators. Among the characteristics of flowers attractive to insects are

1. The production of particular scents.
2. Colour, size, and shape of petals.
3. Patterns of stripes of spots on petals.
4. Whether petals are separated or not.
5. Shape of flower.

Scent production is usually very important in attracting a potential pollinator and often in determining behaviour following landing. Flowers pollinated by true flies, which breed in dung, may actually mimic the odor of dung, so "fly" flowers commonly have a disagreeable odor for humans. On the other hand, "butterfly" and "moth" flowers usually have a sweet smell. Flower colour may also be correlated with the type of pollinating insect. For example, honey bees and their relatives are attracted by many blue, purple. Yellow flowers, but cannot "see" red flowers. However, butterflies are attracted to red flowers. As stated already that many insects (*e.g.*, honey bees) can see ultraviolet light. In fact, many bee-pollinated flowers that appear to be unpatterned solid colours to us turn out to have characteristic stripes or spots when photographed with ultraviolet-sensitive film. These stripes, spots, and so on are visible to honey bees and may serve as visual guides into a flower nectary in addition to simply being attractive to these insects. The shape and depth of nectaries are often such that only particular species are able to extract nectar from them. These specially constructed nectaries also serve to inhibit "undesirable" insects. The amount of nectar secreted by a nectary is such that pollinators get enough to make the effort worthwhile energetically. But not so much that they are satiated by visiting one or even a few flowers.

Examples of the extremes of plant adaptation that have occurred in association with pollinators include imprisonment of pollinating insects and pseudocopulation. Yeo (1973) describes the following fascinating case of imprisonment as a mechanism that culminates in pollination.

In Arum nigrum Schott the spathe opens overnight and the spadix emits a bad smell. In the morning dung-frequenting beetles and flies are attracted by the scent, and alight on the spadix appendage and the inner surface of the spathe. The cells of the epidermis of these organs each bear a downwardly directed papilla. These papillae are of such size that the claws of an insect can grip them, but as the grip trightens the claw is simply forced off the tip of the papilla. The entire surface is covered with an oily secretion which is believed to nullify the effects of the pulvilli. The result is that insects walk a short distance and then suddenly fall. The heavier ones take wing as they fall but the smaller ones are too slow, and drop through the ring of bristle-like sterile florets into the "prison". At this time the stigmas can be pollinated by pollen carried by the incoming insects. The following night the stigmas cease to be receptive; pollen is then shed into the chamber and dusts the insects. By the morning the sterile florets, previously smooth and oily, have become wrinkled so that insects can walk on them. The papillae on the spathe and spadix also shrink, and the insects can escape.

FIG. 2.10: Marsh marigolds as they appear to humans (A) and to insects sensitive to ultraviolet light (B).

Certain orchids (Ophrys and others); produce stimuli that attract their specific pollinators (particularly Diptera or Hymenoptera, by releasing copulatory behaviour. These orchids produce scents that mimic the effects of sex pheromones as well as stimulating the insects visually and tactilely.

The most common insect pollinators are various members of the orders Coleoptera, Lepidoptera, Diptera, and Hymenoptera. Among the common adaptations of pollinating insects are elongated mouth-parts, that facilitate obtaining nectar from flowers with deep nectaries, plumose (featherlike) hairs on the body to which pollen clings, and various specialized pollen-collecting and/or transporting structures such as the corbiculae (pollen baskets); on the hind tibiae of honey bees and their relatives and pollen brushes on the hindlegs of most bees.

Another good example of the evolutionary "fine tuning" that can occur between mutualists is seen in the association between attine ants and their fungus gardens. As ants prepare a fungus garden, they add their own fecal material, which benefits the fungus. The fecal material contains proteolytic enzymes and various nitrogenous materials that play vital roles (*i.e.*, protein digestion and enhancement of growth) in the metabolism of the fungus. In turn, the fungus contributes cellulase, which benefits the ants by catalyzing the digestion of cellulose. Such close metabolic interdependency is characteristic of many mutualistic associations.

Parasites and hosts apparently pass through successive cycles of "parasite attack-new host defense," the parasite tending to become increasingly benign and the host more tolerant, perhaps eventually leading to a commensalistic or even mutualistic outcome. For example, parasites that invade the insect hemocoel are commonly encapsulated. The presence of such a capsule around a parasite may severely limit the parsite's activities by inhibiting mobility or interfering with feeding and/or gaseous exchange. On the other hand, some parasites secrete materials that block capsule formation. Duffey contains information on chemicals in insects that have an antibiotic effect. Whitcomb *et al.*, (1974) consider insect defenses against micro-organisms and parasitoids.

A remarkable example of an extremely close adaptive adjustment between parasite and host is the relationship between the rabbit fleas Spilopsyllus cuniculi and Cediopsylla simplex and their rabbit hosts. The reproductive behaviour and physiology of these fleas is strongly influenced by the rabbit's reproductive hormones. Egg maturation only occurs on pregnant doe rabbits and has been shown to be stimulated by the host's corticosteroids. Estrogens and other hormones also contribute to egg development and influence oviposition. Larvae develop in the rabbit nests in which the young rabbits are born. Linkage of the sex cycles of the fleas and rabbit hosts ensures that the reproductive activity of the fleas occurs such that newborn flea larvae end up in the host's nest.

That predators of insects and their insect prey have exerted strong selection pressures on one another is clearly evident in the means predators have evolved to capture prey and the avoidance mechanisms of potential prey. Several examples of coevolutionary adaptation involving insects as predators and insects as prey have already been given in earlier chapters. Here we consider examples of insect adaptations in response mainly to selection pressures by insectivorous vertebrates, in particular protective coloration, form, and pattern: and co-evolutionary adaptations in bats and moths.

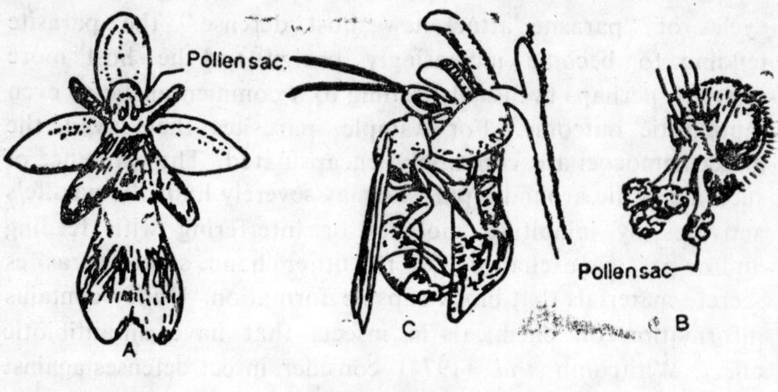

FIG. 2.11: A. The orchid Ophrys insectifera, which visually resembles a female bee. B. A bee attempting to copulate with the flower. C. As the result of attempted copulation, pollen sacs become attached to the bee's head.

Examples of protective form, coloration and pattern iuclude crypsis, aposematic coloration, mimicry, and spots. Such protective devices would be of diminished value (or even deterimental) without appropriate behaviour. Thus, although we will emphasize form, coloration, and pattern in this discussion, it must be remembered that "appropriate" behaviour has also come about as a result of selection pressure applied by predators. General treatments of protective form, coloration, and pattern include Cott (1975), Portmann (1959), and Wickler (1968).

Crypsis refers to various combinations of form, colour, and pattern that facilitate "hiding" from potential predators. Resemblance to various inedible, even repulsive, objects like thorns or bird droppings; coloration and shading that cause an insect to blend into the background; and transparent wings that allow the background to show through all belong in the crypsis category.

A classic example of crypsis (and one that has contributed significantly to our understanding of natural selection) is inustrial melanism. Industrial melanism refers to phenomenon first observed in the peppered moth, Biston betularia, in England. These moths are nocturnally active and during the day they rest on lichen-covered tree trunks. Prior to 1845, collection records of peppered moths described them as light-coloured, an appearance that provided them with camouflage when they rested on the tree trunks. In 1845 and thereafter, dark (melanic) forms began to appear around the industrial centre of Manchester and eventually came to be more common than the light-coloured forms. It was hypothesized that the change in predominant phenotype (from light to dark) was associated with the overall darkening of tree trunks due to industrial pollution. Such darkening was thought to lessen the selective advantage of light-coloured moths while favouring the survival of dark forms. Birds were assumed to be the agents of selection. Kettle well demonstrated experimentally that birds did, in fact, prey more on uncamouflaged forms; light-coloured forms had a definite selective advantage in unpolluted areas. He also showed that moths tended to rest on a substrate that matched their body coloration, indicating that they could discriminate between light and dark backgrounds. With the advent of pollution controls, light moths are beginning to appear again in industrial regions.

Aposematic or warning coloration is sometimes called "advertising" in the sense that an insects very bad tastes (*i.e.*, is unpalatable), stings, or does something similarly disagreeable to a predator and communicates this ability visually by colours and patterns that contrast blatantly with the background. Vertebrate predators soon learn to avoid these insects on the basis of a few unpleasant encounters. An aposematically coloured insect may also be advertising falsely in that although it looks like (mimics) a harmful insect, it is really palatable and harmless. Aposematically coloured insects are usually boldly patterned in shades of orange, red, or yellow contrasting with black and stand out against the greens and browns that the characteristic of the environment. Unpalatability is commonly due to the presence of chemicals in the hemolymph that have been derived from host plants. For example, the monarch butterfly (Danaus plexippus;) and certain true bugs and beetles, which feed on various species of milkweed plants (Asclepiadaceae), accumulate chemicals (specifically cardiac glycosides) from these plants. These chemicals not only render the insects unpalatable to vertebrate predators espicially birds) but also act as emetics (*i.e.*, they make a predator vomit). Although all monarch butterflies are aposematically coloured, an individual may or may not be unpalatable depending on the level of cardiac glycoside in the particular milkweed plant it fed upon as a larve. In an interesting coevolutionary turn-about, certain insectivorous birds have circumvented the monarch's chemical defense by learning to reject or ingest individuals on the basis of taste. In this instance aposematic colouration would tend to be disadvatageous for the insects since it would render them more visible to predators.

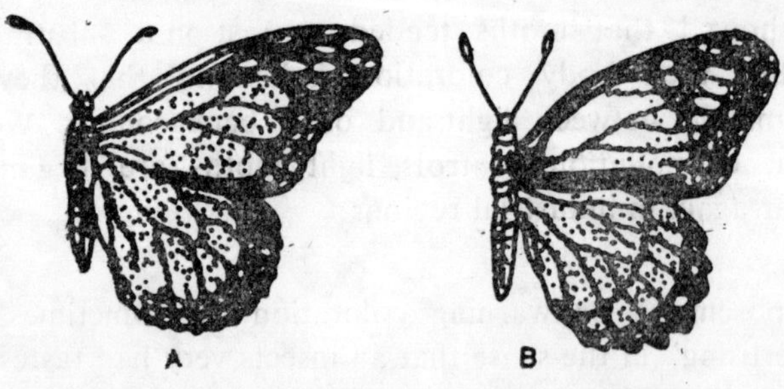

FIG. 2.12: The monarch butterfly. Danaus plexippus (A). serves as the model for its Batesian mimic, the viceroy butterfly Limenitis archippus (B).

Mimicry, as used here, refers to the close resemblance (in form, colour, pattern, or otherwise) between one insect, the mimic, and another insect or a plant part, the model (insects may mimic thorns, leaves, etc.). Various forms of mimicry have been identified in insects, among them Batesian, Mullerian and Wasmannian. A Batesion mimic is palatable (and otherwise harmless), but is protected somewhat by virtue of its similarity to an unpalatable (or otherwise harmful) model. An oft-cited example is the viceroy butterfly (Limenitis archippus), which is a Batesian mimic of the monarch butterfly. Batesian mimics are commonly less numerous than their models. This is interpreted as advantageous on the basis that large numbers of aposematically coloured but palatable individuals would reduce the avoidance learning in predators. Miillerian mimics differ from Batesiian in that both mimics and models are unpalatable (or otherwise harmful). Several species that bear close resemblance to one another may comprise a Mullerian mimicry complex, all members gaining advantage by virtue of the fact that the number of different colours and patterns predators are required to learn to avoid are reduced; that is, predators need learn only one pattern to result in protection for all the insects in the complex. Wasmannian mimics are insects that live in close association with ants and that have come to closely resemble their hosts. In this case the models (*i.e.*, ants) are considered to be the selecting agents, the guests who fail to resemble their hosts sufficiently being attacked.

Many insects (*e.g.*, certain moths, butterflies, and caterpillars) bear spots at various locations on their body. These spots often closely resemble eyes of vertebrates and perhaps, in some instances, eyes of other insects. Many moths, which have eye spots on the hindwings, are otherwise cryptically coloured when resting on their usual substrate. If the first line of defense, crypsis, fails and a predator strikes, the forewings are thrown forward, suddenly displaying the eyespots. Such behaviour has been shown experimentally to frighten and repel insectivorous birds. The eyespots resemble eyes of birds (*e.g.*, owls) that prey on the insectivorous birds. Small spots are commonly found along the edges of the wings of many butterflies. These have been shown to serve as pecking targets and by virtue of their location away from vital parts of the body may provide protection by diverting the peak of a predator and allowing time for escape.

A superb example of predator/prey coevolution is the association between insectivorous bats and members of certain families of night-flying moths, in particular Noctuedae, Arctiidae, and Geometridae. Bats locate prey by emitting pulses of sound beyond the range of human sensitivity (ultrasonic). As these sounds bounce off various objects, living and nonliving, the bats hear the echo and use this information to guide their flight and to locate flying prey insects. Details of the discovery and elucidation of this echo-locating ability are recounted in Griffin (1959). Tympanic organs that perceive the echo-locating cries of bats have evolved in several moth families. The moths display various escape maneuvers in response to bat cries. The coevolutionary "battle" has proceeded a step further in certain Arctiid moths, which possess microtymbal organs on the metathorax. When a bat approaches one of these moths, trains of sound pulses are generated by the microtymbal organs. These sounds have been shown to, in effect, "jam" the bat's echo-locating system and hence interfere with its ability to catch the moth. Herbivorous insects and plants have played major roles in one another's evolutionary history. Several structural and behavioural adaptations of herbivorous insects are described elsewhere.

FIG. 2.13: Io moth (female). Automeris io, with forewings thrown forward, displaing eyespots.

Plants have responded, in the evolutionary sense, in a variety of ways to the continued onslaught of herbivorous insects. Among the adaptations that provide defense against insect herbivores are numerous structural and biological characteristics, but the major defensive adaptations involve chemicals.

Structural characteristics that interfere with insect attack (feeding and/or oviposition) include tough cuticle, hard seed coast, velvetlike pubescence, and a variety of spines and thorns. Recently certain plants have been found to be covered, at least in part, with highly specialized, minute, sharp-pointed, hooked hairs (hooked trichomes). These hairs hook into the integument of caterpillars (Heliconius spp.) feeding on Passiflora adenopoda or leafhoppers feeding on certain strains of field beans immobilizing and puncturing these insects, which eventually succumb to starvation and desiccation. More information on trichomes may be found in Levin (1973) and Rathcke and poole (1975).

Limited growing seasons, nutrient-poor sap (relative to herbivorous insects), and mechanisms that allow rapid dispersal of seeds (which increases difficulty in food location for seed eaters) are examples of various biological means by which plants "combat" herbivores.

Many chemicals found in plants have no apparent role in plant metabolism, but are present in comparatively concentrated amounts. In several instances, these secondary plant substances (secondary metabolic products), have been shown to be of value-to the plants by influencing the behaviour or physiology of herbivorous insects. That is, they sometimes act as allomones.

Secondary plant substances may have toxic, repellent, deterrent, or hormonal effects on insects. The toxic effects of certain plants have long been known, and several botanical insecticides still find use today. Examples include nicotine (derived mainly from Nicotiana tabacum, the common tobacco plant), the pyrethrins (from various species of Chrysanthemum), and rotenone (From Derris spp. and many other legumes). Citronella and cedat oil are examples of insect repellents derived from plants. Deterrents of various sort have been identified. These may be substances that are ingested and influence feeding or such things as resins that physically interfere with feeding.

Substances with molting-hormone actively have been found in several ferns (Polypodiaceae) and two families of gymnosperms. Many different phytoecdysones have been isolated and analyzed; the most common is β-ecdysone, the major insect-molting hormone. Many phytoecdysones are more potent than insect hormones. These substances protect plants by interfering with insect development. Juvenile-harmone activity has also been identified in certain plants especially balsam fir and eastern hemlock. Juvabione, a juvenile-hormone mimic from balsam fir, is very similar in chemical structure to juvenile hormone. Like the phytoecdysones, juvabione and other plant substances with juvenile-hormone activity cause pathological effects, such as inhibition of metamorphosis and sterilization of adults. Juvenile-hormone activity in plants was discovered by chance when the bug Pyrrhocoris apterus failed to complete development in cages the bottoms of which were covered with a particular kind of paper towelling. The hormonally active paper towelling turned out to be produced from balsam fir. Recently, a secondary plant substance with antijuvenile hormone activity has been identified from the plant Ageratum houstonianum.

Many insects have "fought back" against plant defenses. An insect (or any hervivore) that breaks through a given plant's defenses obtains a double advantage; if finds a new source of food and does not face competition from other herbivores that have not broken through the plant's defenses. One mechanism that has provided some insects with protection against secondary plant substances has been the development of microsomal mixed-function oxidases (MFOs) secreted in the midgut. These enzymes degrade a wide variety of substances; for example, they are active against pyrethrins. However, in the case of Chrysanthemum (producer of pyrethrins) the plant has responded with the production of a substance called sesamin, an MFO inhibitor. Interestingly MFO systems are best developed in polyphagous insect species, which are most likely to be contronted by a variety of secondary plant substances.

Some insects not only have broken through plant defenses but also have come to use secondary plant substances to their own advantage. For example, as previously mentioned, the monarch butterfly accumulates cardiac glycosides, which render this insect unpalatable to predators. For other insects secondary plant substances serve as kairomones, that is, as attractants that identify food plants.

3

Defensive Mechanism in Insects

Insects and Food for Vertebrate Animals

Vertebrate animals are commonly classified as herbivores, omnivores, or carnivores. Since strict herbivores are themselves primary consumers, their main associations with insects are as competitors or as hosts (for example, cattle grubs). Many omnivores and carnivores consume insects at least some of the time, and some are mainly or wholly entomophagous (or insectivorous) for example, bats, swallows). Many freshwater fish depend heavily on insets, not only those occurring in the water but those that fall in or fly low over the surface. This is, of course, well known to fly fishermen, who pride themselves o designing flies that resemble certain types of insects and on striking the surface film at the right time and place with the appropriate fly. Fish will commonly continue to feed for a time on a type of prey that is abundant, for example, a flight of newly emerged caddisflies. Trout tend to take food both at the surface and from the bottom, mayflies, caddisflies, and midge larvae being major items in the diet. The productivity of lakes and streams, in terms of energy made available by producers and primary consumers, is a major concern of fisheries biologists.

Mosquito fish (Gambusia) have been introduced into many parts of the world for control of mosquito larvae, often with much success. To turn to another group the Amphibia, the giant toad has been widely used in warner parts of the globe to control ground-dwelling insects. Insects make up a considerable portion of the diet of many amphibians and reptiles.

Many mammals also subsist on insects. Even large carnivores such as the coyote consume a great many grasshoppers and beetles. Two whole orders, the Chiroptera (bats) and Insectivora (moles and shrews), are largely insectivorous. Studies in central New York by W.J. Hamilton, Jr., of Cornell University, showed that short-tailed shrews consume food equivalent to 50 per cent of their own weight each day, and about 75 per cent of their diet consists of insects. Surprisingly, deer mice, rodents one commonly thinks of as plant feeders, also consume about as high a proportion of insects during the summer. In Canada, shrews and voles are major predators on a major predators, on a major forest pest, the larch saw fly. In some areas 37 per cent to 93 per cent of the cocoons are eaten by these mammals, and a portion of the surviving adults and their larvae are eaten by birds. It is, of course, birds that have attracted the most attention as predators on insects. Several books have been written on the value of birds, often without regard for the fat that they consume many species that are beneficial to man Being active fliers, birds do, however, have the capacity to congregate in areas of insect outbreaks, as evidenced by starlings flocking to grub—infested lawns, or warblers concentrating on trees infested with caterpillars. Perhaps the best-known instance of birds destroying noxiousinsects relates to gulls and the Mormon cricket. About 1850, great swarms of these crickets descended on Mormon settlements near Great Salt Lake, destroying most of their crops and threatening the settlers with starvation. Gulls appeared by the tens of thousands and devoured most to the crickets, an historical event of such importance that a statue to the gulls was later erected in Salt Lake City.

Although many birds are rather generalized feeders, most have some degree of specialization. Warblers and vireos, for example, forage for small insects among the leaves and small branches of trees, while blackbirds and thrushes forage mainly on or close to the ground. Flycatchers take flying insects they snatch in quick flights from their perches, while woodpeckers drill for the larvae of bark beetles and other wood-infesting insects. It is not always appreciated that many seed-feeders (such as finches) consume large numbers of insects during the summer season.

The numbers of insects consumed by individual birds are very large indeed. Most birds fill their crops with food about twice a day. There is one record of a flicker stomach containing over 5000 ants, another of a captive robin consuming 165 cutworms in one day, nearly twice its own weight. Nestlings commonly consume more than their own weight in a day, the food in many cases being a hig-protein diet of soft-bodies insects. Studies of a nesting pair of blue tits, in England, showed that the 11 nestlings were fed from 25 to 42 times an hour, from early in the morning to late evening, for a total of 400 to 650 visits per day. The number or weight of insects required to sustain an individual bird through its lifespan would be difficult to calculate, but the figure would obviously be very high.

It should be kept in mind that, as a general rule, the feeding behaviour of vertebrate animals differs considerably from the of insects. Insects tend to be programmed to respond to only a few of the innumerable potential stimuli in the world around them, for example, Nasonia wasps to fly puparia, horse bots to the legs and flanks of horses. Vertebrates are more broadly responsive and may sample a variety of foods before selecting those they prefer and learning where to find and how to catch those that suit them best. Many birds are known to develop a search image; this is, they discover a particularly edible item, remember its appearance and where to find it, and hunt in such a way that others of the same kind are quickly discovered. Experiments have shown that birds will concentrate on one type of prey to the point that is numbers have been reduced far below those of other, equally palatable species, which have been overlooked. At a certain point a new search image will, of course, be formed. Birds also become conditioned to avoid prey that sting, taste bad, or make them ill. It is not surprising that insects have developed a variety of adaptations that take advantage of these characteristics of vertebrate predators.

THE PRIMARY DEFENSES OF INSECTS

It is convenient to recognize two major types of defense mechanisms. Primary defenses operate before a predator initiates an attack, and in fact regardless of whether or not a predator is present. They may also be thought of as passive defenses, in the sense that the insect is, by its appearance and actions, merely bearing a message to potential predators. Secondary defenses are employed at the time of an encounter with a predator; they are active in that the insect has to behave in some way *vis-a-vis* its attacker. An insect may have both primary and secondary defenses. So diverse are these mechanisms that we can do no more than mention a few well-studied examples here.

Crypsis

Crypsis is a widespread phenomenon among insects. Although often called "camouflage" or "protective coloration," in fact it implies more than that. To be cryptic (which literally means "hidden"), an insect must not only resemble its substrate, but it must also behave appropriately, for example, by resting immobile or in an appropriate posture. Generalized crypsis implies an overall resemblance to the background, for example, by a speckled grasshopper resting on a pebbly surface Special resemblance implies similarity to a specific object, such as a twig or a leaf. An insect may resemble its abiotic environment, as in the case of the grasshopper cited above. Or resemblance may be to the biotic environment, usually some part of a plant, and may vary from the simple green colour of a caterpillar in one's salad to bizarre body forms copying lichens, spines, or even flowers.

Some insects are able to undergo an enhancement of crypsis by assuming more than one colour or by altering the environment to render themselves less easy to detect. Some mantids and grasshoppers undergo colour changes when they molt, and in some cases these are induced by humidity. High humidity may cause changes to green colours, low humidity to brown, depending on the colour of the immediate environment. This is believed to be a direct effect of light reflected from the substrate. Examples of insects that modify their environment are provided by the larvae of certain leaf beetles, which cover their backs with cast skins, fecal pellets, and bits of debris, thus resembling from above nothing more than masses of detritus.

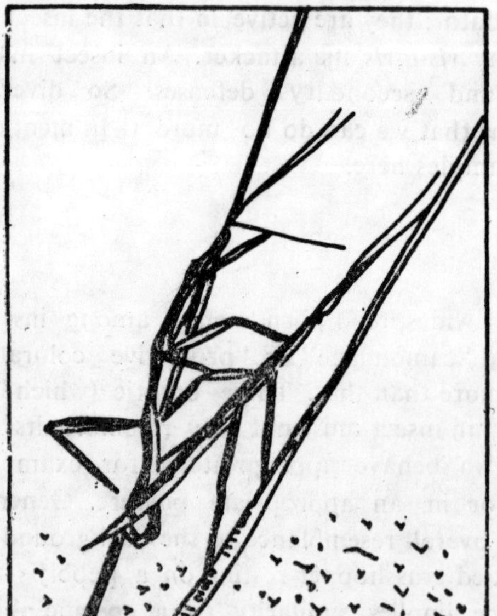

FIG. 3.1: A pair of mating walkingsticks, the male on the left.

Cryptic insects must have behavioural responses that function to orient them suitably. Bark-dwelling moths with longitudinal stripes on their wings must orient up and down on striped blark, while those with transverse stripes must position themselves sideways if the stripes are to blend with patterns in the bark. Certain caterpillars that infest pine needles are green with longitudinal white stripes, rendering them difficult to observe when resting on the needles. But at the last molt they become brown, flecked with black, and from then on they become brown, flecked with black, and from then on they rest not on the needles but on the brown twigs bearing the needles.

Crypsis does, of course, interfere with other life activities, such as freeding and mating. Thus it is especially prevalent during the diurnal resting periods of insects that are mainly active at night, especially moths. Cryptic insects do not usually aggregate, for birds are frequently able to form search images for cryptic species, and by dispersing themselves these insects increase their chances that birds will find more rewarding prey. There are varying degrees of crypsis among species, and among individuals of many one species some are more successful than others in escaping detection. It is these better-adapted individuals that, on the average, survive to produce more offspring, thus providing the basis for the evolution of more perfect crypsis. At the same time a premium is placed on the keenest hunters among the predators. Thus there is coevolution between predator and prey, resulting in improved abilities to find prey and more refined examples of crypsis.

The Effectiveness of Crypsis

It is true that examination of the stomach contents of various birds, lizards, and mammals often reveals the presence of cryp ic species. However, there is now much experimental evidence that increasing degrees of crypsis result in increasing success in escaping detection by predators. For example, laboratory studies using chameleons as predators have been conducted with a species of grasshopper having two colour forms, green and yellow. Appreciably fewer green than yellow grasshoppers were eaten when placed on a green background, and *vice versa*.

Perhaps the best—known studies are those of H.B.D. Kettlewell, of Oxford University, concerning a common British insect, the prepared moth. These moths are whitish, with irregular black spots and stripes, and blend beautifully with lichen—covered bark. Before the industrial revolution, one occasionally found black (melanic) individuals that were conspicuous against the lickens. But by 1895, around industrial

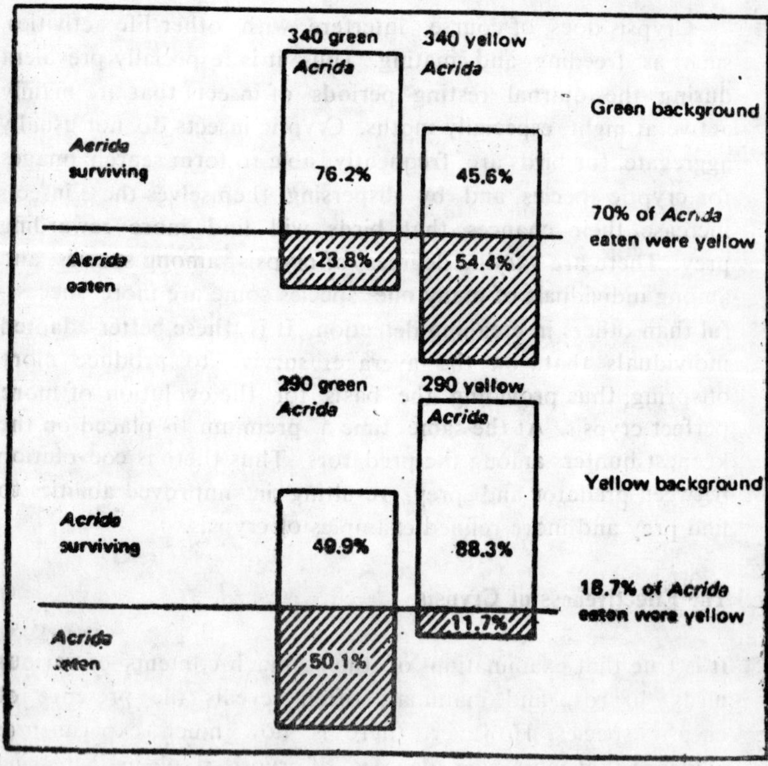

FIG. 3.2: Predation by a chameleon on green and yellow grasshoppers of the same species (Acrida turrita) on a green background (below).

centres in England, the melanics had become much more abundant than the "normal" form. By this time tree trunks in these areas had become blackened with soot and the lichens killed off; thus the melanics blended best with the background. Kettlewell hypothesized that birds preyed more heavily on the whitish forms in industrial areas and thus had produced rapid selection toward melanism (known to be controlled by a single dominant gene). In order to demonstrate this, Kettlewell reared large numbers of normal and melanic moths in the laboratory, marked them with coloured spots, and released them in both industrial and rural areas. These were then recaptured in light traps or traps baited with pheromone. In industrial areas, many more of the melanics than normal moths were recovered, and in rural, unpolluted areas the reverse was true. He also placed moths of both types on tree trunks in selected areas and observed them from blinds. In one rural area, where tree trunks were lichen covered, 190 of the moths placed on the trunks were eaten by birds within a short time; 164 of these were dark and 26 light. Obviously mohs that look conspicuous to us also appear conspicuous to birds. In recent years various pollution control programmes have been instituted in England, and, as might be predicted, the melanic moths now comprise a smaller percentage of peppered moth populations in industrial areas than they formerly did. These moths have onlyone generation a year, so it is obvious that natural selection, in the form of predation, may provide a powerful force in the evolution of crypsis.

Aposematism

Aposematism is a general term for signals that advertise unpleasant on dangerous attributes of an animal. The term warning coloration is often applied to aposematic features, but this term imlies more volition than is appropriate and overlooks the fact that features other than colour may be involved. Aposematic insects all have secondary defense mechanisms, such as a sting or distasteful or poisonous body fluids. Predators must either have innate avoidance responses to aposematic patterns or must learn these patterns by sampling prey. In the latter (probably much more common) instance, some individuals of each generation of an aposematic insect are likely to be sacrificed. However, some species have evolved tough bodies that cannot easily be damaged or "deflection marks" that induce the predator to bite in some nonvital part of the body.

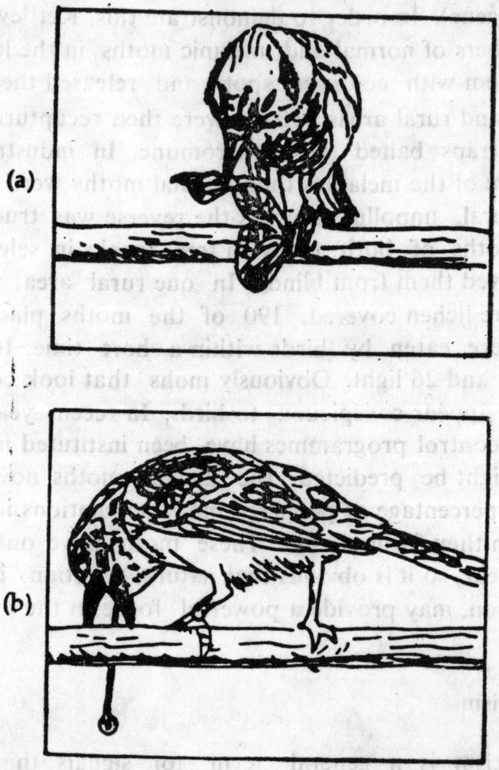

FIG. 3.3: A blue jay (a) feeding on a monarch butterfly and (b) becoming ill within a few moments and vomitting. (Lincoln P. Brower, 1969, "Ecological Chemistry," Scientific American, 220: 2, pp. 22-29. Reprinted by permission).

One of the best—known aposematic patterns consists of alternating bands of black and yellow, the pattern of yellow jackets and many other stringing Hymenoptera. These bright, alternating bands are conspicuous to colour—blind persons, and one assumes also to colour—blind animals (such as many mammals). A second common aposematic pattern consists of orange or red bands or patches, which are especially effective signals to organisms with good colour vision (which is why hunters and highway crews wear orange jackets). Birds, in general, have colour vision similar to ours, and it is probable that these patterns evolved primarily as signals to birds. One example that quickly comes to mind is to monarch butterfly, the larvae of which feeds on milkweek. Most milkweeds contain cardiac glycosides that are retained in the bodies of caterpillars, the pupae, and on into the adult butterflies; when a bird eats a monarch it becomes ill and retches.

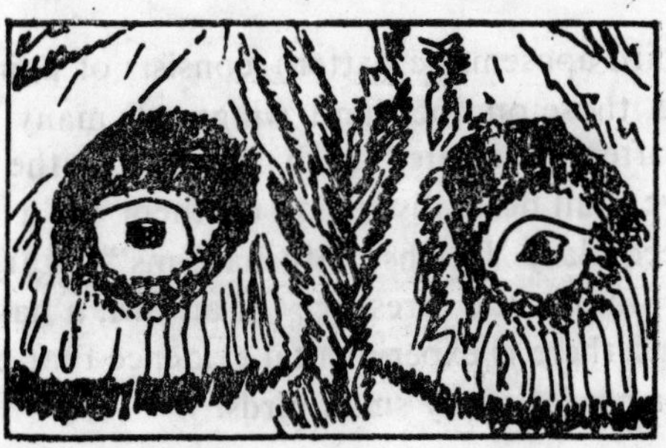

FIG. 3.4: The eye spots on the hind wings of a polyphemus moth, one of the giant silk moths (Saturniidae). When a resting moth is disturbed, the front wings are spread, suddenly revealing the eye spots.

Another example is provided by wingless female multillid wasps ("velvet ants"). These insects spend much time walking over the ground in search of their hosts. Although a few species are cryptic, and coloured like the sand, the majority advertise their presence with bands of brilliant orange and red. With their long, venom—laden stings and tough integument, it is not likely that many are actually sacrificed in the learning processes of predators.

It should be remembered that colour vision in insects differs somewhat from that of humans and higher vertebrates, being extended on the blue end of the spectrum and into he ultraviolet and not extended appreciably into the oranges and reds. It is true that some insects use bright colours in nuptial displays (damselflies, for instance), but the colours employed are usually blues (even ultraviolet in the case of some butterflies). It is no coincidence that flowers requiring insect pollination are usually blue or yellow, or have ultraviolet nectar guides; or that flowers that rely on humming—birds for pollination (such as scarlet gilia) are primarily red. Insects displaying orange or red marking almost without exception have secondary defense mechanisms and are advertising this fact to potential vertebrate predators—or they are mimics of such insects, as we shall discuss in the following section of this chapter.

Still a third aposematic pattern consists of paired eye spots, for example, those on the hind wings of many moths. Since these are often associated with displays in the presence of a predator, we shall defer discussion of them to a later section (under "Secondary Defense Mechanisms"). Large eye spots presumably imply the presence of an owl, a hawk, or a large mammal, and there is experimental evidence that eye spots do in fact deter predation by small birds.

Before leaving the subject of aposematism, it should be noted that it is common for many insects in a given area to have similar aposematic patterns. In temperate regions a great many different kinds of stinging Hymenoptera, for example, have "yellow jacket" patterns. In the tropics there are often literally dozens of butterflies in the same area having similar orange wing patterns, many of these feed on poisonous or bitter—tasting plants and are themselves unpalatable. This phenomenon was first noted in Brazil by the German biologist Fritz Muller and has since been termed Mullerian mimicry, although we prefer the term mutual aposematism. Muller pointed out that it is advantageous for distasteful species occurring together to evolve similar patterns, since in this way predators have fewer patterns to learn and exploratory predation is spread over several species.

In contrast to cryptic species, those displaying aposematic patterns tend to occur in expose situations and are often gregarious. Larvae of the mourning cloak butterfly provide a good example of gregarious feeders, covered as they are with long spines and streaked with red; but the chrysalids are cryptic, and prior to pupation the caterpillars crawl off in different directions, crypsis being enhanced by dispersal.

The Effective of Aposematism

There is now much evidence that aposematism "works." In a pioneering study, the British entomologist G.D.H. Carpenter offered a variety of insects to monkeys and found that 73 per cent of the cryptic insects were eaten but only 20 per cent of those with bright colours. Lady beetles, with their characteristic red and black patterns, are known to be avoided by many kinds of birds and mammals. The British entomologists J.F.D. Frazer and Miriam Rothschild "screened" a great number of insects for palatability, using caged animals, including diverse birds and mammals, lizards, and toads. Lady beetles fell in the highest (most unacceptable) category. Equally unacceptable was the cinnabar, a lepidopteran that has been widely used in biological control of ragwort. The caterpillars of this species are gregarious and are striped with black and yellow, while the adults have red and black wings. The blood of the cinnabar contains considerable quantities of histamine as well as certain alkaloids. In nature, cinnabar caterpillars are preyed on extensively by ground beetles and are parasitized by braconid wasps, but shrews and voles will not accept them.

One of the best—known field studies on the adaptive value of aposematism was conducted by W.W. Benson, of the University of Washington. His studies were conducted in Costa Rica on a well-known, unpalatable butterfly, Heliconius erato, having a bright red band on the black fore wings. Benson captured butterflies in their gregarious, nocturnal roosts and painted out the red bands with black; in the controls he merely painted black over black, to check against any effect the paint itself might have. By recording the return of these butterflies to the roost, he found that the experimentals survived an average of 32 days, the controls an average of 52 days. Furthermore a much higher percentage of the experimentals had wing damage resulting from predation by birds, even though they were exposed for a shorter period, on the average, than the controls. It should be added that several other butterflies have much the same pattern as Heliconius erato, one of many examples of mutual aposematism in the American tropics.

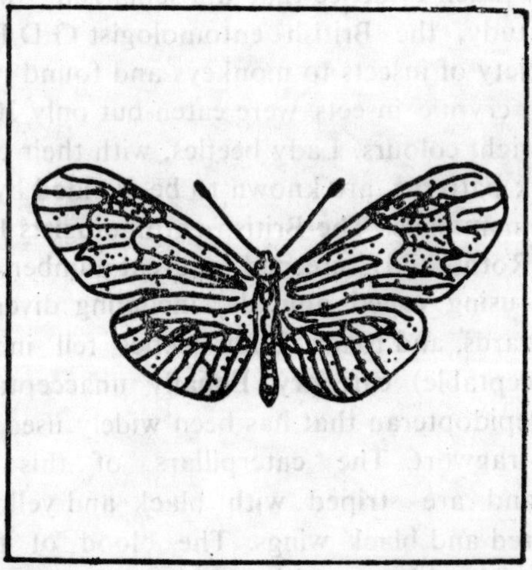

FIG. 3.5: Heliconius erato, a tropical American butterfly considered highly unpalatable. The front wings have a bright red band, the hind wings a yellow streak.

The Evolution Mimicry

Many similar experiments have now been performed with diverse mimics and predators, with generally similar results. Some of the most interesting ones have confronted one of the major criticisms of mimicry theory: how can one account for the evolution of (for example) a victory from a black and white ancestral species without invoking a sudden saltation or total genetic reorganization? In other words, how can mimicry evolve when being "just a little bit mimetic" would seem to have no possible advantage? Lincoln Brower and his associates at Amherst College have addressed this problem, using Heliconius erato, mentioned above in another context. They found that birds trained to avoid this butterfly also refused to take uniformly black butterflies of similar shape as well as red—banded butterflies of a different shape. In further studies, they varied the unpalatability of the model and found that the more unpalatable it was, the more the birds tended to extend their rejection to species only rather remotely resembling the model. Stimulus generalization such as these experiments imply means that under certain conditions even a slight resemblance to an unacceptable model may have value in terms of natural selection. If in each generation individuals with most resemblance to the model leave even a slightly greater number of progeny than average, in time a more "perfect" mimic may evolve.

	Experimentals (n=26)	control (n=42)	
	11	6	
Percentage of insects with using damage	42.3%	14.3%	× 2=5.31P<0.05 (1)
Total days of exposure to predation	888	2103	
Days exposure to predation per insect	34.2	50.0	

FIG. 3.6: Wing damage to Heliconius erato experimentals (red painted over with black) and controls (black painted over previously black areas). Controls also survived much longer, thus were exposed to predation for significantly longer periods.

It should be noted that in many butterflies only the female sex is mimetic, the male having the "normal" colour of its group. In at least some cases, colours of the male play a role in courtship and mating, and it is advantageous for males to retain patterns that release sexual behaviour in the females. The tiger swallow—tail of the eastern United States, rovides a particularly interesting case. The females of this butterfly are dimorphic, having either mainly black or yellow and black "tiger" coloration. Males always have the tiger pattern and are most successful in mating with females coloured like themselves. But in the southeastern states there is a mostly black butterfully that has been shown to be highly distasteful to birds: the pipevine swallow tail, In the South—East, tiger swallowtail females that are black mimics of this species survive longer, on the average, but yellow and black individuals have greater mating success. Thus a balanced polymorphism is retained. But in the North, where the pipevine swallowtail does not occur, all females are "tigers", since here there is no selective advantage in being black.

The diana fritillary, of the south-eastern states, exhibits similar sex—limited mimicry, the females copying the pattern of the pipevine swallowtail, the males retaining the usual speckled pattern of fritillaries. In this case there is no polymorphism; but case are known, especially among African swallowtails and relatives of the monarch, in which there are not two but several different female morphs, each having a colour pattern like that of a different distasteful model. In situations involving polymorphism, the mimetic species may achieve a population size greater than that of its model, since each morph has a different model. In general, it is disadvantageous for mimics to become more plentiful than their models, since birds will encounter more mimics than models and learn much more slowly to associate distastefulness with a particular pattern. Before leaving this subject, it should be mentioned that mimics must also behave like their models to be effective, just as a cryptic insect must "behave like a piece of bark", or whatever it resembles. Cases are known in which mimicry is primarily behavioural. For example, certain species of darkling beetles have glands at the tip of their abdomen from which they are able to spray repellent benzoquinones. When approached by a predator, these beetles assume a characteristic stance, with the abdomen high in the air, ready to discharge their secretions. Black ground—dwelling beetles of at least two other groups also assume this stance when approached by a predator, even though they lack these glands.

Aggressive Resemblance

Since insects so readily evolve one or more of several cryptic, aposematic, or mimetic patterns, it is not surprising that some predatory insects have evolved coloration or behaviour like that of their hosts or have evolved crypsis serving primarily to gain access to a host. Such behaviour hardly qualifies as "defense"; rather it is offense. We discuss it briefly here because it has often been termed "aggressive mimicry", though it represents quite a different phenomenon from true mimicry, not only as to function but as to the nature of the signals: for these are, of course, directed towards other insects and not toward vertebrate animals.

Some of the best examples of aggressive resemblance occur among the many thousands of inquilines in the nests of ants and termites. Many of these, beetles especially, are remarkably ant like in form, and some produce secretions that are eagerly sought by their hosts and may resemble pheromones of their hosts. Many termitophiles and myrmecophiles produce or acquire cuticular odours that copy those of their hosts. These inquilins commonly feed on food stuffs in the nest and some actually feed on immature ants or termites.

A most remarkable instance of aggressive resemblance occurs among fireflies and has been described by James Lloyd. of the University of Florida. Male fireflies produce a pattern of flashes characteristic of their species, and it is answered at a specific interval by females of their species on the ground.

Females of certain species have evolved the ability to respond to flashes of males of alien species. These males fly down to the females and are promptly consumed.

SECONDARY DEFENSE MECHANISMS

Many insects have structures or behaviour effective at close quarters with a predator. This applies not only to aposematic species, all of which are assumed to have undesirable qualities, but to many palatable insects. In its simplest form, it may simply involve a butterfly dashing away from a pursuing bird, or a caterpillar thrashing violently when seized by a predator. Some of the more specialized defense are striking indeed, and we shall be more particular concerned with these.

Flight Patterns

For small, flying insects the best defense may be escape. Probabilities of escape may be increased by swift or evasive flight or by an abrupt colour change on settling. Band – winged grasshoppers show brilliant colours of the hind wings while in flight, but when they land, these are covered by the cryptically coloured front wings, causing the insect to "disappear" suddenly. Many butterflies (anglewings, in particular) have the upper wing surfaces brightly coloured, but when they land only the lower surfaces are exposed, and these are commonly coloured like bark or dead leaves. It is believed that flash colours such as these promote escape from pursuing birds.

Bats are major predators on night—flying moths. Several groups of moths have evolved receptors turned to the ultrasonic cries that bats use in the manner of sonar in locating prey. These receptors have only two sensory neurons each, one for low—intensity sounds and one for high—intensity (closer) sounds, and since they are bilateral, the moth has means of perceiving both distance and direction. Moths respond to bats at some distance by flying swiftly away, but at close quarters they undertake erratic flight patterns or sudden power dives.

One group of moths, the tiger moths, has gone one step further and evolved sound-producing organs that advertise to the bat their distasteful qualities. These moths respond to the approach of a bat by producing a series of clicks, and it has been demonstrated that bats veer away from clicking moths.

Death Feigning

Since many predators are attracted to moving prey and reject dead insects, it is not surprising that many relatively defenseless insects become inert when approached. Leaf beetles and weevils are especially prone to death feigning. This is easily observed by some gardeners who control Colorado potato beetles by hand—packing; for as soon as disturbed, these beetles draw their legs beneath them and drop to the ground. As soon as the disturbance is over, they climb back onto the plant and resume their feeding.

Spines, Poisonous Hairs, and Stings

Many caterpillars are hairy, and some are covered with stiff, branched spines. In some cases the tips of the hairs or spines break off easily and are capable of causing momentary irritation or a rash. This is true, for example, of the densely spiny larva of the io moth. The sting of Hymenoptera is believed to have evolved primarily as a means of paralyzing the prey, but in bees, many ants, and social wasps it has lost this function and serves in defense. The nature of the venom of social species differs considerably from that of solitary forms, consisting of a mixture of pain—producing substances. In a few species (such as the honey bee) the sting is barbed and remains imbedded in the skin, causing the death of the worker and releasing a pheromone that may bring other workers to the attack. It is interesting that the males of several groups of solitary wasps (which lack an ovipositor and therefore a sting) have the last abdominal segment prolonged into a stout spine, which is wielded much like a sting and can sometimes pierce the skin, though there are no poison glands. It is probable that such a "pseudosting" is nearly as effective as a true sting in deterring predators, especially since male wasps commonly have aposematic patterns similar to those of the females.

Detachable Body Parts

Many insects have integumentary outgrowths that readily become detached, without seriously harming the insect. These include the waxy or powdery coverings of certain aphids and whiteflies, the hairs of caddisflies, and the scales of bristletails and of moths and butterflies. Thomas Eisner and his colleagues at Cornell University have shown that moths and caddisflies are often held only momentarily by spider webs, since the scales or hairs stick to the adhesive strands and become detached from the body, permitting the insects to escape with minor damage. In a similar way, these insects may escape from the viscid droplets on the leaves of sundews. To what extent deciduous outgrowths play a role in escape from vertebrate predators is uncertain, but it is known that moths shed clouds of scales when pecked by birds. It is probable that scales first evolved as an escape mechanism and only later, in many Lepidoptera, assumed a role in conveying messages via colour patterns. That is, an originally secondary defense may have become also a primary defense, reducing the frequency of predator attacks and at the same time retaining a role in escaping from such attacks.

Deflection of Attack

Many butterflies have small spots along the edge of the wing, and it is believed that these attract the attention of predators and cause them to bite at a nonessential part of the body. Butterfly collectors are aware that he wings of fritillaries and others having such deflection marks on the margin of the wing commonly have triangular beak mark at the margin. The "tails" of swallowtails may also serve a role in deflecting attack from the body, and as collectors are all too aware, these are commonly damaged unless the botterfly has recently emerged. Hair—streaks have not only antenna like filaments at the posterior margins of the wings, but usually a colour spot at the base of the filaments, simulating a head. Thus a perched hair-streak often seems to have its "head in the wrong place," causing a predator to seize only a piece of the wing.

FIG. 3.7: Sticky strand of a spider's web. A moth has escaped from the web, leaving some of its scales behind.

Startle Displays

Some insects, when approached closely or attacked by a predator, suddenly undergo movements, produce sounds or scents, or display colours serving to "threaten" or "bluff" a predator. The usual effect is probably to startle or to cause a momentary indecision, permitting the insect to escape. There is evidence that birds and mammals remember encounters with displaying insects and to a certain extent avoid them. Both cryptic and aposematic insects may have startle displays as a "second line of defense" in the event of attack.

Walkingsticks, which one thinks of as the ultimate in crypsis, have a variety of displays. Many tropical species have short, brightly coloured hind wings that are suddenly erected, abruptly increasing their size; this sometime may be accompanied by stridulation that is said to resemble the hissing of a snake. Michael robinson, of the Smithsonian Tropical Research Institute, in Panama, has described a number of such cases, including one wingless species that flexes its abdomen up and down while stridulating, suggesting a scorpion or (more remotely) a snake about the strike. Some walking sticks have sharp spines capable of pricking an aggressor, and others have caustic secretions that can actually be aimed at an aggressor. A number of different insects produce squeaking sounds when seized, such as long horned beetles and mutillid wasps. Even the relatively inert, unprotected pupae of some Lepidoptera produce squeaks when handled. Some bright—coloured, distasteful tiger moths produce an odorous froth from thoracic glands when disturbed. Larvae of swallowtail butterflies have eversible hornlike glands on the thorax called osmetoria. These produce a volatile substance smelling rather like rancid butter, and in fact its major component is butyric acid. Swallowtail caterpillars often have aposematic patterns, and many have eye spots that are dramatically displayed by elevating and inflating the thorax.

Eye spots deserve special mention, since they occur in a wide variety of insects and are usually associated with some type of display. We refer here not to the small, marginal deflection spots on butterfly wings, but to the large eye spots often found on the heads, or thoraces of caterpillars or the hind wings of mantids, moths, and other insects. Hawk moths provide a particularly find example. Many species are cryptically coloured but have brightly coloured hind wings, each containing a large eye spot. When disturbed, the moth spread its fore wings and undergoes quivering motions, presumably serving to further startle a predator. Some moths and butterflies have spots shaded in such a way as to have remarkably resemblance to the eyes of owls or snakes. David Blest, of University College, London, has shown that the sudden appearance of any bright colour releases escape responses in certain birds. This proved to be particularly true of eye spots. Blest fed several species of birds with mealworms, using a box that would flash a light beneath, revealing a pattern on the floor of the box. The sudden flash of a pair of parallel lines or crosses caused some escape responses, but pairs of circles causes many more such responses. Eye spots shaded so as to appear three—dimensional were especially effective in "frightening" the birds. He concluded that many small birds respond innately to the sudden appearance of eyes that might represent those of a large predator. However, Blest showed that birds do sometimes habituate to butterflies with eye spots when these prove palatable. Thus eye spots are most effective if not too widespread in nature and when displayed only under certain conditions.

Chemical Defense

The ultimate form of defense is the use of one or more chemicals that are in some way repugnant to a predator. These may be obtained from the hot plant (as in the casre of the monarch butterfly) or synthesized by the insect. Defensive chemicals (called allomones) may be contained in the blood or may be produced be specialized exocrine glands. Many insects that exhibit chemicals defense are aposematically coloured, and some are mimicked by insects lacking such defense, as we have seen.

Integrated Defense Systems

Many insects have not one but several defense mechanisms. In this way they may achieve protection against different predators of against the same predator at different levels of motivation or different stages in the learning process. Commonly larvae and adults are protected quite differently; for example, the larva of the viceroy butterfly resembles a bird dropping, while the pupa resembles a dried leaf, and the adult is a mimic of a distasteful species, the monarch. In the same life stage, insects may have several "lines of defense." Walkingsticks are cryptic, but if attacked they may have startle displays or discharge irritating chemicals at the intruder. Some tiger moths have cryptic fore wings and brilliantly coloured hind wings, which provide flash colours, but if actually attacked discharge noxious fluids from thoracic glands.

In a now classic study of 15 species of preying mantids on Trinidad, Jocelyn Crane, of the New York Zoological Society, found that all 15 were cryptic, some being sticklike, others leaflike, some like bark or lichens. All 15 had escape mechanisms: dropping, dodging, or jumping. As a third line of defense, several species had startle displays, differing in details but usually involving body swaying, spreading the wings, and rising the forelegs and abdomen. As a last resort several species would actually strike the predator with the raptorial forelegs. Crane hypothesized that the elaborate displays evolved by ritualization of simultaneous to remain motionless and to escape.

FIG. 3.8: The startle display of a preying mantid. This mantid is cryptically coloured, but when disturbed it assumes an aggressive pose and threatens to strike, at the same time displaying bright colours on the hind wings.

Perhaps the most impressive integrated defenses occur in the colonies of social insects. Some tropical social wasps produce sounds by drumming on the carton of the nest; and if approached, they display by wagging their brightly coloured abdomens or, as a last resort, stinging the intruder en masse. Yet their defense against army ants may be often have walls nearly as hard as stone, but if the walls are breached, a horde of specialists in chemical defense quickly arrives on the scene.

The question is often asked: if these things are true, why is it that some insects have several defense mechanisms, some have only one, and some appear to have none at all? Apartial answer is to be found in the reproductive capacities. Insects such as aphids or house flies, which have no spectacular defense stragies (but like most insects may have subtle ones), compensate by having usually high reproductive rates. They are specialists in reproduction rather than survival. Every species may be said to maintain itself in nature by achieving some short of balance between reproductive potential and survival capacity.

4

Space-Related Behaviour in Aquatic Animals

Space-related behaviour is associated with the use of space. It is one of the most important determinants of social behaviour in animals, including man. Dispersal is the phenomenon of spacing out or spreading into available habitats. It results from various kinds of behaviour with different underlying motivations. For example, crowding causes an increase in agonistic behaviour, and subordinate or defeated individuals disperse into adjacent habitats. These are frequently less optimal, and so the fitness of the animal forced to disperse is reduced. In such cases, dispersal and its attendant behaviour can be considered as forced in the proximate causal sense of the word. In other cases, dispersal is voluntary in the ultimate evolutionary sense. Animals disperse or migrate into habitats to feed, to reproduce, and to escape from predators. The time frame for dispersal is variable. It may be on a diel, lunar, or seasonal basis. In all cases, however, the resultant behavioural and ecological measure of dispersal is dispersion. Since dispersion is the static pattern of distribution of animals in space, it determines their proximity to one another and therefore affects the type, frequency, and intensity of social interactions. This is true for behaviour both within and between species. Clearly, then, an appreciation of space-related behaviour in animals is essential for an understanding of their social behaviour. Recognition of the applicability of these ideas to the understanding of human behaviour is evidenced by the emergence of what may be called biosocial anthropology and human ethology.

Since the utilization of space affects important demographic parameters such as population density and distribution, it indirectly affects the breeding systems and hence the gene low and age structure of populations. These are important determinants of social organization. Thus, a thorough understanding of the sociobiology of any species must include a careful study of its space-related behavioural.

In this chapter we shall discuss the study of space-related behaviour in aquatic animals, to explore the intuitive, pervasive ideas underlying the persuasive concept of territoriality, and to analyze selected examples of space-related behaviour in fishes and compare them with the solution to similar problems by terrestrial animals, especially birds and mammals.

SPECIAL PROBLEMS AND METHODS OF STUDY

There are special problems associated with studying the space-related behaviour of aquatic animals. They are due to the facts that instrumentation is more difficult to employ under water, and, perhaps most important, that the human observer—still the most sophisticated data collector—is severely limited in the time that he can spent underwater, for physiological reasons and, in the case of saturation diving, physiological reasons and the costs of operating an underwater habitat. Of course, certain studies can be done in aquaria, but even here the major restriction is space, which makes it doubly difficult to feel confident that the experiment is testing the variables under question—namely, the use of space. This isn't to say that excellent work on space-related behaviour cannot be done in aquaria. It certainly can, and Van den Assem's experimental study of territoriality in the three-spined stickleback is an outstanding example.

Electronic Tagging

Besides conventional tagging and recapture techniques that provide information on gross movements, ultrasonic and radio tags are also effective. They provide the possibility of obtaining details on movements during the time the tagged animal is monitored. In addition to the gross movement patterns of large-mouth bass, Warden and Lorio measured eight movement and activity variables. Examples are number of home areas selected, average distance travelled to home areas, and average time spent at home areas. They defined home areas as any location in which bass remained for more than two days. The maximum distance moved from a central point within the home area was used as an indication of home range. Home range size differed with seasonal water temperature. It rarely exceeded 30 m during summer months, but was larger, approximately 75 to 100 m during the spring and fall. The animals were inactive during winter. These data were obtained by the use of 16 tagged animals. Tags had an average life of 1308 days. Tracking was continued daily for approximately 2.2 hours over a 6 week period, providing a total of 570 hours of tracking time.

Inspite of the fact that the electronic lagging and tracking methods are the advance methods to trace out the space related behaviour in aquatic animals, they provide very little information on details of social behaviour. For example, even continuous tracking of an ultrasonically tagged animal does not provide information on other animals that are encountered (unless they are other tagged individuals) or the nature of the encounters. For example, was the encountered animal preyed on, attacked, fled from, courted, or simply ignored: Was it a conspecific? Was it larger or smaller, male or female? If not a conspecific, what species was it? Answers to questions such as these are essential it an understanding of the significance of the movement is desired. Only visual observations of the events can provide answers to these questions. There are simply no alternatives. Thus, if one wishes to carry out a thorough ethological study of an aquatic animal in the field, it is absolutely necessary to observe the animal's behaviour for relatively long periods of time underwater.

Underwater Television

Television provides a powerful tool for underwater observation. It has been used successfully to study the behaviour of two small, territorial pomacentrid fishes that remain in one place on the reef: Eupomacentrus partitus, by Myrberg (1922a, b) and Stevenson (1922), and E. planifrons by Thresher (1975a, b). They combined acoustical studies with their video observations. Given an elaborate and expensive television system, it is possible to keep individual fishes under continuous surveillance for many hours, and thereby to observe the details of their behaviour. Myrberg *et. al.*, (1969) have also used a video-acoustic system to study the behaviour of sharks.

The major limitation to underwater television is that it is stationary, and even when the camera can be tilted and rotated by remote control, the area of underwater topography that is seen by the camera is relatively small. In situations where the daily or seasonal migrations are well known, it is possible for the observer to predict a location from which certain kinds of behaviours can be observed. Under these conditions, underwater television can be used to great advantage.

SCUBA Techniques

There are several species that move over larger areas of the reef in the course of their daily activities. A rough rule of thumb is that the larger the size of the fish, the larger is its area of activity. This is especially true for large predators with the exception of ambush predators. To observe effectively the behaviour of these species, it is necessary to follow the animals underwater, using SCUBA at a discrete distance. Various data-recording devices, such as underwater tape and event recorders and cameras, may be used. Underwater observation time, however, is the most important limiting factor. It becomes severely limiting as depth increases, and, in fact, the traditional observational techniques of ethologists become increasingly impractical at depths below about 10 and are impossible below 20 m. Ethological studies of species living at depths below 20 m can be done only with saturation diving from an underwater habitat. The difficulties of cost and of diver-observer fatigue and psychology are mentioned above.

In butterfly fishes the following three basic underwater observational techniques have been practiced using SCUBA.

1. The first is to observe one area of the reef for a pre-determined period of time. This technique provides detailed information on the behaviour of the species that are resident in the particular location. It is possible to learn individual fishes of certain species on the basis of variation in their marking patterns, old scars, and torn fins. If one is working at depths of 10 m or less, it is possible to carry out three hours of observation a day. It is exhausting, however, and the total time of observation is equal to about half a day's work for an ethologist studying a terrestrial species. Clearly, in many instances, underwater television would be a superior alternative.
2. The second method is to swim transacts over selected areas of reef topography for known distances and predetermined time periods. This technique permits sampling of many different individuals but provides few details of their behaviour. Fishes usually are not recog-nized as individuals. The data take the form of instan-taneous or "point-in-time" observations of distribution, abundance and behaviour. There are no alternative techniques available.
3. The third technique is the most productive for the underwater ethologist, is to select an individual, pair or group of fishes and to follow it at a discreet distance, observing and recording the details of behaviour. The

use of an underwater tape recorder, used either with voice or with a touch-tone type of even recorder, provides a realtime frame for observations. Again, there are no alternatives available. This technique is supposed to be he most fruitful technique, it is also the most difficult and potentially the most dangerous. Dangerous because the fish may lead the observer into deeper water, down current, and at a considerable distance from the boat. The swim back on the surface, without air and with the burden of equipment, can be exhausting. Difficult because the observer must learn what the discreet distance is for the particular species being studies. This takes experience and a gradually acquired intuitive feeling for when the observer's presence affects the behaviour of the fish. In areas where there is much spear fishing, none of these techniques, especially the last one, are effective.

Methods of Mapping

It becomes possible to map the areas over which the fish forage by the effective use of the third technique. Mapping is easily achieved with the use of brightly painted and numbered lead weights, which are dropped at the points of greatest excursion. Distances between lead weights are easily measured after the observation period has elapsed. Preliminary studies, ascertaining that the weights do not affect the animal's behaviour, must be made. For example, if the weights are inspected repeatedly or nipped at, then they should not be used. Since there are limitations on the amount of time that an observer can spend underwater, it is difficult to apply G.K. Noble's classical criterion of defense of an area as the distinction between home range and territory. With only a few hours of observation daily, it is unreasonable to assume that defense of an area has in fact occurred, or, conversely, not occurred. The problem is confounded further if one consider the nuances and subtleties of threat behaviour that, taken alone, may suffice as active defense of an area. Furthermore, since fishes as a group undoubtedly are able to learn each other as individuals, brief bouts of agonistic behaviour, perhaps days apart, may serve the purpose of area defense. The chances of observing such bouts increase directly with total observation time.

For these reasons, the terminology and methodology developed by Weeden in her study of territorial behaviour of the tree sparrow are especially appropriate to the study of space-related behaviour of aquatic animals. Weeden argues that it is possible to measure accurately the space used by an animal in a defined time interval, and this measure is more useful to the behavioural ecologist than is an inaccurately measured defended area. The method is best applied with intensive and repeated observations of a few animals in a relatively small area. Daily activity sapce and total activity space are geographical units combining all kinds of observed behaviours in an arbitrarily defined time interval and time of day. Daily activity space in Weeden's study was based on daily 4 hour observation periods. At depths of 10 m or less and in warm tropical waters, 3 hours of observations, in two or three consecutive dives with short rest periods in between, are not unreasonable for an underwater ethologist. Total activity sapce is a composite of daily activity space. The technique developed by Odum and Kuenzler (1955) is used for determining when enough observations have been made for a particular animal. Since dataon behavioured activity are recorded on a prepared grid sheet representing the study area, grids with different amounts of activity can be identified, and in this way the total activity space can be represented as a series of concentric circles representing a core area of intense utilization, an area of moderate behavioural activity, and an outer "daughnut" of space of lowest use but with its outer boundary including the entire total activity space.

A somewhat similar method for investigating area utilization within the home range was developed by Tester and Siniff in their study of raccoons. It was used successfully by Ables in his study of the home range of red foxes. An interesting comparison of techniques used to measure the area of activity of terrestrial animals is given by Mohr and Stumpf.

TERRITORIALITY

Territoriality is only one point on the behavioural scale of space-related behaviour, but it has received more attention than all the other categories combined. The subject has been repeatedly reviewed, and recent general treatments are to be found in Brown, Brown and Orians, Stokes and Willson. Interesting examples of territorial and home-range behaviour in mammals are reported in Ewer, Jewell and Loisos, and Geist and Walther. In birds, space-related.

Behaviour, especially territoriality, has been the focus of much work. Besides the general reviews given above, Orians and Willson (1964) examine interspecific territoriality. Schoener (1968) extensively reviews sizes of feeding territories, and Cody (1974) views territory in the framework of community structure. Many of the ideas that follow have been gleaned from a perusal of these reviews.

Stocks' (1974) book entitled Territory deserves special mention. To read the collection of papers he presents is both an instructive as well as a humbling experience. Humbling because one learns that many of our current ideas about territoriality, including the concept of cost and benefit analysis, were recognized 120 years ago, in 1868, by Bernard Altum, and that critical questions regarding territorial behaviour, such as the role of bird song, were recognized 90 years ago, in 1903 by C.B. Moffat. And Maragaret M. Nice initiated careful quantitative studies on the movements and territories of song sparrow 52 years ago, in 1937. Since these pioneering studies, much has been written on the subject in both the primary and the secondary source literature, and much of this literature is reviewed in the selection of recent references given above. Form a sampling of this vast literature there emerges the "intuitive concept" of territoriality.

The Intuitive Concept

The intuitive concept of territoriality includes the defense, if necessary, of a particular place and the prior, if not exclusive,

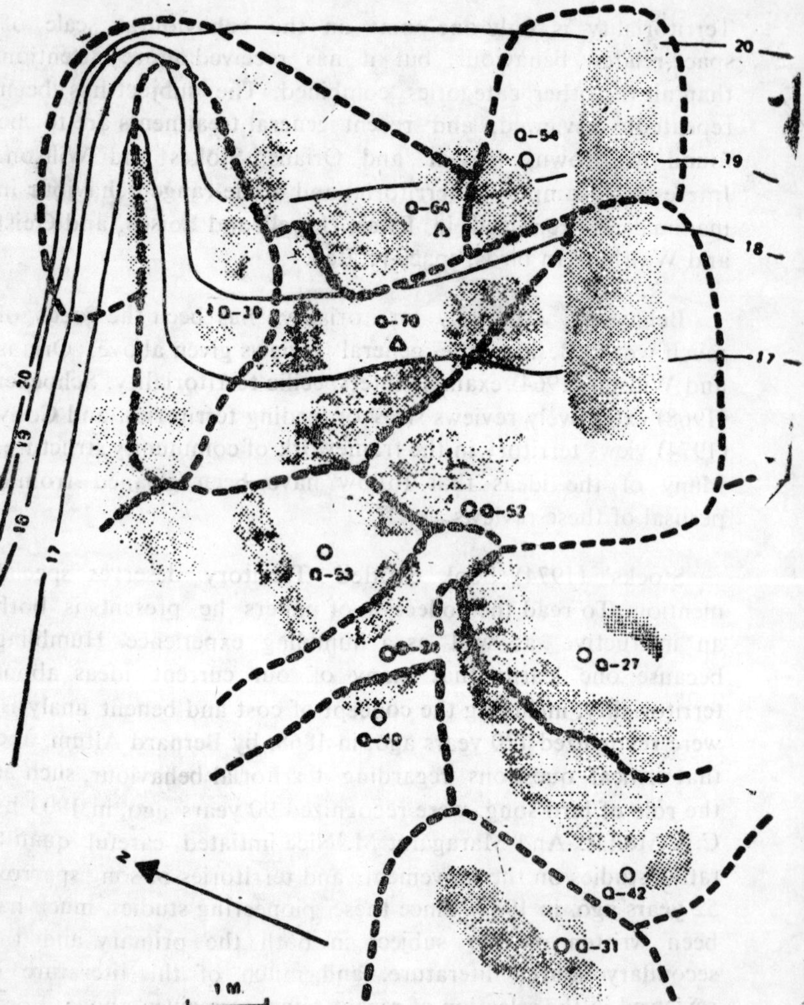

FIG. 4.1: A. Territories of 11 Hypsypops rubicunda. Dashed lines indicate the limits of movement during a 15 min observation period. Nest sites of nine nesting males are shown by circles; shelter holes of the remaining two individuals are shown by triangles. Contours are in meters. Large rocks are shaded.

use of the resources therein. The emphasis should be one use. On the face of it, this does not seem to be a particularly complex system, but when the time dimension is added and the nuances of complex social behaviour, such as individual recognition, are superimposed, then ways in which territoriality may be expressed effectively become very complex and defy simple definition. For example, let us examine the idea of the exclusive use of space. Multiple use of the same space may occur if the uses are separated in time, or if the resource is not space *per se*, and the two users are utilizing different resources in the same space. The observer would see two or more users in the space and conclude that it was multiple use when, in fact, each user would be utilizing the space exclusively for itself. The situation is relatively simple if use is separated by time. The same refuse may be used by a nocturnal and diurnal species. For example, reef fishes generally sleep on the bottom, in holes or crevices on the reef. Although supportive data are needed, it is postulated that the same refuges on the reef are used by different species on a day and night basis.

The same type of food may be eaten by different individuals of the same or different species if eating periods are separated in time, and if different portions of the food resource are selected. Given sufficient food distributed in a relatively large area, it's possible to divide the food resource by using the territory exclusively at different points in time and by consuming different sizes of food. This system still works for large predators, such as leopards and tigers, which differ in the size of prey killed, the use of vegetation types, and the time of activity. The same resource sharing can be expected to occur in large aquatic predators such as sharks and barracudas. However, if food becomes scarce, a condition necessitating extended foraging movements over greater distances and periods of time, then the probability of maintaining the exclusiveness of the use of the area decreases. Tension encouters and agonistic behaviour may result, under the guise of territorial defense. The question then arises: Whose territory? The answer will be forthcoming when dispersal occurs. The situation can arise when food and space are shared by two different species, when they are both either nocturnal or diurnal; but if one of the species is diurnal and the other nocturnal, then the situation should not arise. However, food shortages may result in changes in the diel pattern of foraging. Thus, the concept of prior or exclusive use, at first relatively simple and straightforward, becomes increasingly complex. Nevertheless, territoriality is best thought of in terms of the benefits of resource use rather than the costs of territorial defense. The above considerations apply equally well to the concept of home range.

How space-related behaviour involving the prior or exclusive use of the resources within a particular space increases fitness is particularly challenging to the underwhat ethologist. In this same regard, it is important to understand why one species is territorial while another is not, or why one species forages over the reef in male and female pairs, as do a number of omnivorous species of butterflyfishes while other species do not.

Effect of Topography

All animals, including man, centre their activities on important topographical features. Pioneering trade settlements were built on rivers and fortresses on high ground. Baboon troops have sleeping trees in favoured locations within their territories. The presence of a shrub or tree in which to build a nest is a prerequisite of many bird territories. Obviously, there is a close relationship between space-related social systems in animals and their environment, both physical as well as biotic. Since space-related behaviour must be adapted to the habitat in which it is expressed, it is to be expected that the shape of territorities and home ranges will be influenced by topography.

Along rivers, streamp, and lake sloves an interesting parallel exists between the linear home ranges and territories of butterflyfishes and those of a number of species of mammals, including muskrats, otters, and wolves. The muskrat is a semiaquatic rodent, and it prefers to establish itself along shorelines with adequate vegetation and with little wave action or current. Tracts of stream edge, lake shore, or marsh that are recognizable as territories may be as far apart as 1000 m in sparsely occupied habitats, but at higher densities they are separated by only 20 to 40 m. Females, not males, are primarily involved with territorial defense, especially when they have young. Aggression is directed toward other species, such as mink and conspecific strangers. Local resident individuals are tolerated. Consequently, it is again difficult to characterize muskrats as either territorial or home ranging. They show both kinds of space-related behaviour, depending on the situation. Erlinge found that family groups of fish otters consisting of a mother and her young cubs establish their home ranges along streams and occasionally the shoreline of lakes. The size of the home range, which may be as large as 7 kms, is affected more by the growth of the cubs, topography, and access to open water in winter, than by food supply and population density. Adult male dog otters wander over greater distances and may have home ranges as large as 15 kms. In winter, otters become territorial. The families of the river otter range about 3 to 10 km in a season, and about three families occur in an area along 20 kms of stream, as noted by Lievs.

Several species of mammals like wolves live in packs, and packs space themselves in the wild in areas of hundreds of square miles. According to mech, it is not clear whether packs live in true territories in the sense of a defended areas, because meetings between packs in the wild have rarely been seen. The few known cases are of large packs meeting single, or pairs of, strangers. Intolerance was shown by the large pack. They chased the strangers and if they caught them attacked them. Mech observed fleeing lone wolves run at top speed for as long as a mile after the pack had stopped chasing them. The evidence suggests that territorial spacing does occur, and packs probably avoid one another. There is relatively little overlap in the use of space by smaller packs, and, if it does occur, the packs use the space at different times. Movement of packs often, but not always, follows topographic features, such as rivers, ridges and valleys. Burkholder found moose kills along rivers and creeks, whereas caribou kills were not so distributed.

In red foxes the shapes of the home ranges are influenced, but not determined entirely, by habitat features. The same is true of mice. In arctic ground squirrels, territories are necessary for successful breeding and for hibernation, but again topography is of great importance. Subordinate, younger, less experienced animals establish territories in marginal areas and are subject to greater mortality from drowning and predation. However, Kruuk found that natural obstacles or topography are not important in establishing boundaries for home ranges and territories of the spotted hyena. This finding supports the suggestion of Stumpf and Mohr that linearity of home ranges and territories, which they believe is a characteristic of many mammals, may be due to behavioural factors not associated with habitat terrain.

Among invertebrates there are many interesting examples that support the opposing view—namely, that habitat topography does play an important role in determining the shapes of activity areas of animals. Male ghost crabs defend the areas around their burrows, where they construct a display pyramid. The burrows are arranged linearly within the intertidal zone of sandy beaches. Dragonfly males establish reproductive territories along the shorelines of rivers, streams, and lakes, and a prerequisite of the territory is that it contain a suitable perch, usually a shrub leaf or twig, where the male can settle and rest after courtship display. In many coral reef environments, a relatively flat shallow water area extends out from the shore to a variable distance where the bottom drops off at a steeper grade into deeper water. At Heron Island on Australia's Great Barrier Reef, the drop-off has an estimated grade of between 20° to 40°, and it is here that one finds the most luxurious growth of corals and the greatest concentration of coral reef fishes.

Pairs of a number of species of omnivorous butterflyfishes, including Chaetodon ephippium, C. unimaculatus, and C.

vagabundus, maintain what appears to be extended, linear home ranges along the upper portion of the reef slope. The same is true for C. auriga, which may occur in pairs but more often is seen swimming alone or in small groups along the reef front. Interestingly, it tends to swim in pairs at Enewetak Atoll in the Marshall Islands (Reese). Since individually recognized pairs often were not seen for a number of days, either in the study area or along the transect area, which extended 100 m along the reef slope, the home ranges of these species are believed to exceed 100 m in linear dimention. Furthermore, the pairs were seen within a depth of 10 m and never more than perhaps 5 to 10 m from the surface of the reef. Thus, the home ranges are visualized as long, thin, narrow volumes along the reef slope. The distinction between home range and territory in these species is based on the single criterion of whether or not bouts of agonistic behaviour occurred when the pairs met. Unfortunately, the sophisticated criterion of use, as discussed above, was not applied during the study. Fighting occurred only between pairs of C. ephippium, and hence their living space may be referred to as an intraspecific territory. In any case, both within and between the species, living spaces overlapped as different pairs were observed at different times in the same area of the reef. Avoidance behaviour, in the sense of "not paying any attention to one another" or just slightly altering swimming direction so as to avoid one another, seems to be the most common form of social interaction, both intra-as well as interspecifically. Chaetodon ephippium is the exception. Pairs do show aggressive behaviour toward each other when they meet, both at Heron Island and on the reefs at Enewetak Atoll in the Marshall Islands. Recent observations by L. Boucher at Enewetak indicate that the social group of C. unimaculatus is larger than a pair and that subtle bouts of agonistic behaviour occur between the groups. MacDonald observed a linear arrangement of male reproductive territories of the pomacentrid Microspathodon chrysurus along the reef drop-off in Puerto Rico.

It has been observed that habitat topography affects the shape of home ranges and territories of mammals, arthropods, and coral reef fishes. Further study of this relationship is called for, since it is clear that there is a similar responses in the space-related behaviour of these diverse animals to a particular dimension, linearity, of their habitat.

CENTRAL PLACE SYSTEMS AND TRADITION

Refuging

Refuging is considered only as a central place system of space utilization on a diel rhythm for the procurement of food. However, the concept is applicable in a larger time frame to lunar and seasonal movements related to feeding and reproduction. The concept has little utility, however, when applied to single events in an animal's biology, such as the spawning migrations of salmon.

Generally, it should be looked for where food resources are distributed widely in an area surrounding a particular part of the environment in which the animals take shelter or find refuge when not feeding. Perhaps the best-known example is that of starlings returning to a central roosting location to spend the night after a day of foraging for food as individuals or in smaller social groups. The same phenomenon is well known for mynah birds in Hawaii, where the raucous noise they make in their roosting site, often a large banyan tree, can be annoying. Hamilton and Watt give other examples from birds, mammals including man, and social insects, ants, and bees. Presumably there is a benefit to be accrued from the behaviour of coming together in one central place. It may be related to defense from predators, and it may serve functions of social priming.

Now a days the refuging behaviour is becoming well documented for fishes as well. Many hervivorous and omnivorous species migrate along frequently well-established routes from nighttime sleeping refuges to day-time feeding grounds. Nocturnal species, for the most part predators, show a reverse migration, often moving considerable distances offshore to feed. The grunt, Hoemulon flavolineatum, spends the day hiding in holes on the reef, but at dusk it aggregates, and within 20 to 30 min after dusk it streams out, following well-defined routes, on to the surrounding seagrass flats where it feeds until dawn. At dawn the movement is reversed, and they return to their home reef for the day. Large mixed schools of acanthurids swim from one area of the reef to another to feed and may be exhibiting refuging behaviour. In all probability planktivorous species show refuging behaviour as they move in groups from their shelters on the reef to feed in the water column above. Sharks are known to form large aggregations during the daytime that presumably disband at night for individual foraging. Barlow described the daytime aggregations as the refuging groups.

Radial Packs

Central place systems seem to have evolved in response to the coaction of increasing population size relative to the distribution pattern of resources, usually food, making the exclusive use of fixed space too costly for the benefits derived therefrom. In the aquatic environment, the food resources of plankton and algae are frequently widely distributed, and animals feeding on them may be expected to show refuging behaviour.

A derivation of refuging as a central place system is the occurrence of radial packs. Instead of individuals dispersing from the central place refuge, larger social groups either roam throughout the peripheral area or move into specific home ranges or territories. A prerequisite here is some degree of integrity and organization over time of the dispersing social groups. Whether or not radial packs occur in fishes is a challenging question for underwater ethologists.

Tradition

Tradition is the passing of specific farms of behaviour from generation to generation by learning. The predictable movements of groups of fishes along well-defined migration routes, often over considerable distances, either on a diet or a seasonal temporal basis, suggests the possibility of tradition. Tradition is well known in birds and mammals, especially primates.

The traditional learning of seasonal home ranges and the migratory access routes to them by mountain sheep (Geist) is an outstanding example of tradition and one that lends itself as a model to the study of tradition in fishes. Since fisher are capable of associative learning, all that is needed for the study of tradition is the occurrence of different age classes of individuals migrating together. The younger individuals could then learn the traditional route from the older, experienced individuals. Although Willson examines this question on the basis of seasonal spawning migrations of fishes, he concludes that tradition may not exist in fishes. The question is a challenging one for underwater ethologists. In other aquatic animals, tradition most certainly exists in the well-known migrations of marine mammals where the young accompany their mothers in small social groups that may in fact be extended family groups.

PLANKTON FEEDING

Plakton are the free floating, mostly microscopic aquatic a masive organisms including the photo synthetic and heterotrophic farms. Planktonic feeders are those feedigg on these forms.

The feeding of fishes on plankton above a reef is a particularly interesting example of space-related behaviour. There are a number of important differences between plankton feeding and other kinds of feeding behaviour. In most cases, the predatory or grazing animal moves over a two-dimensional surface in search of food. In contrast, the planktonic food resource is distributed unevenly in patches in three-dimensional space. Furthermore, currents carry the food to the consumer, making it possible to remain stationary. Thus, the foraging strategy of certain plankivorous fishes is reduced to the maintaining of a particularly advantageous position in space. From this position, the fish makes visually directed darting movements at planktonic prey as the prey is swept in the current past the fish's feeding position. Stevenson called this characteristic feeding movement "deflection." It involves orienting oneself and moving toward the prey, visually inspecting it, and engulfing or rejecting the food item. Since the prey is coming "down stream," there exists the possibility of olfactory discrimination as well. The strategy is most closely approximated by ambushing predators, but, even in these cases, rarely is the food presented in three-dimensional space. An exception is spiders, which construct their webs so as to intercept airborne prey. Bats and certain insectivorous birds demonstrate feeding behaviour that in many ways is similar to plankton feeding, but the distinction to be made here is that the planktivorous predator feeding above a reef remains in one position, and the prey is transported to it.

The planktonic food resource is variable in space and time, but the group of planktivorous fishes, maintaining itself in space above the reef by swimming into the current, is behaving differently from a foraging flock of birds. The end result, the effective capture of food, is the same, but the behaviours involved are markedly different in their deployment. Unlike foraging flocks of birds, groups of planktivorous fishes feeding in the water column above a reef are stationary. What is crucial to an individual fish is the success with which it can maintain itself effectively in a particularly advantageous position from where its prey-catching success is greater than that of its neighbours. There may be further subdivision of space, with the group's feeding area based on competition and dominance. Myrberg found this to occur in the pomacentrid, Eupomacentrus partitus. The same considerations are relevant to maintaining a particular position within a school of pelagic planktivorous or small piscivorous fishes. Such schools forage in the open sea for their prey, and there is greater similarity to foraging flocks of birds.

In plankton feeding potentially has components characteristic of at least three of Krebs' strategies—hunting by expectation, area-restricted searching, and niche hunting—and may well also involve hunting by search image, although Krebs questions the usefulness of the search image hypothesis. Eggers presents a model for prey selection by planktivorous fishes. The model seems overly simple from the behavioural point of view, since it ignores intraspecific social factors such as dominance relationships, the effects of predators, and possible evasive behaviours of zooplankters. Nevertheless, it is a testable model that should be applied to inshore reef fishes as well as pelagic planktivores. Motivational factors acting on the planktivore, such as the potential threat of a predator and its distance from shelter, may override the factors of prey density, patchiness, capture success, and so forth, on which the model is based. Similarly, as Eggers correctly recognizes, the energy expenditure to maximize prey capture may exceed the benefits in terms of net energy gain. Hobson and Chess point out that predation pressures from planktvores have influenced the evolution of shallow water plankters.

Certain locations are more advantageous than others. A feeding location where the current is sufficiently strong to carry in a continuous supply of food must be balanced against the current being so strong that the must expend a large amount of energy to swim against it. If the space above the shelter is viewed as a series of concentric homispheres, then the further distant the feeding location is from shelter, the larger the potential feeding area can be, and, presumably, the fewer competitors there will be. However, these benefits are offset by the increasing potential risk from predators. Submarine light and underwater visibility are confounding variables, because the detection of both prey and predators is visually mediated. Studies of the escape response in the damselfish Chromis cyanea to crude model that simulated a predator showed that speed of approach of the model and its shape were unimportant, but larger, darker models elicited escape behaviour at greater distances. Cues in addition to these visual stimuli are suspected. It is exceedingly interesting to note that habituation did not occur at intertrial intervals of 0.5 to 4 min. The adaptive significance of this result is obvious if one visualizes the repeated attention a predator may give to a group of prey until it is successful in capturing one or a few prey. Eibl-Eibesfeldt noted that C. cyanea always escape to the same refuge in a predictable way that suggests familiarily with the landmarks in its immediate spatial environment.

The visually mediated escape response of a small cyprinid fishes has been studied by Dill. He found that the reactive distance for flight to occur from both real and artificial predators could be predicted from three parameters: predator approach velocity, predator size in terms of front diameter, and the rate of change of visual angle. It is interesting that escape velocity, although higher in response to real predators than to artificial ones, is not correlated with reactive distance. Furthermore, the reactive field is circular, indicating that the escape response is not dependent on binocular vision. Seeing the predator approach with one eye is sufficient. Naive fish show the response. Reactive or flight distance and escape velocity increase with experience, but the angle of escape is not affected. There is no decrement in the learned change in reactive distance, even after 10 days without reinforcement. It is clear from the studies by Hurley and Hartline and Dill that the escape response is a highly adaptive behavioural mechanism that is the result of strong selection pressure for antipredator behaviour in the two species of fish. There is every reason to believe that selection pressure for an escape response is just as strong in other planktivorous fishes (Chromis cyanea is a planktivore) that leave the shelter of the reef to feed in the water column above it. Therefore, the potential presence of a predator and the distance from shelter are important parameters that must be included in any analysis of plankton-feeding strategies. The success of individuals at different feeding locations could be measured by collecting them after extended bouts of feeding and examining their stomach contents both in terms of quality and quantity of food ingested. Long-term observational data may provide evidence of the costs, in terms of predation, of occupying peripheral, more distant, feeding locations as opposed to locations closer to shelter. The size of the feeding area necessary to maintain a fish of a certain size will be directly proportional to the abundance and distribution of plankton, its nutritional value, and the rate at which the plankton is carried into the feeding area by the current.

The places where plankton is scarce and distributed in widely scattered patches, and the current velocity is low, the planktivorous fishes must have a large feeding area and must make large deflections in order to get enough to eat. Under these conditions, feeding location is of great importance. Where food is abundant, patches are close together, and the current is flowing rapidly, the planktivore may be able to obtain enough food with only minor deflections or movements from its feeding location. The position of the feeding location in the water column is less important under these conditions. The model becomes more complex when conspecific or heterospecific food competitors are added, and, of course, when predators are added. Intraspecific competition would tend to result in individuals feeding further and further away from their shelter, and in the feeding area getting larger, in much the same way as, in theory, the area under the niche exploitation curve increases with intraspecific competition. However, larger feeding areas more remote from shelter increase the risk of predation. Presumably, then, subordinate individuals would be forced to occupy these more dangerous feeding locations. The extent to which the same argument holds or interspecific competitors depends on the narrowness of the feeding guild to which they both belong. Both types of competition may lead to separation of the feeding guild. This separation is expected to occur with interspecific competition as suggested by Root. The separation may occur within a species when different-size classes of individuals exploit different portions of the food resource.

Thus, we see that is many ways, plankton feeding is a unique type of space-related behaviour that deserves further study. It is becoming increasingly evident what the critical parameters are that must be measured in order to demonstrate how this interesting and specialized type of feeding behaviour affects fitness, and it is a challenge to look for comparable situations in terrestrial animals.

5

Avian Behaviour

Many behaviours in birds are inherited and thus innate or "instinctive"—they are performed without preliminary experience or learning. Innate behaviours normally occur in patterns—fixed action patterns – so remarkably constant that they must be triggered by stimuli and can be discharged only after being activitated by internal releasing mechanisms selectively responsive to the stimuli.

The stimuli triggering a particular innate behaviour are special features of the environment—form, colour patterns, movements, sounds etc.—that function as cues and are called releasers. Practically all bird species have social releasers frequently termed signals or displays. These may be visible or audible, and their principal and often exclusive function is to broadcast information to felllow members of the species.

Two near-classic examples of social releasers, proved by experiments, are the black malar strip (mustache) of the male Yellow-shafted Flicker (Colaptes auratus) that releases courtship behaviour in the female and attack by an intruding male and the red patch on the lower mandible of the Herring Gull (Larus argentatus) the releases food-begging in the chick. Both examples are derived from markings that stand out in contrast to the adjacent colour and are thus conspicuous. Other examples of releasers that send out stimuli are particular movements and positioning of the head and/or other parts of the bird's body, vocalizations such as songs and alarm cells, and so on.

Through experimentation, Tinbergen discovered that he could produce super-normal stimuli that would elicit a more effective response. For instance, he found that an Eurasian Oystercatcher (Haematopus ostralegus), which normally lays three, sometimes four, eggs, prefers five if given a choice between three and five, and if offered a gaint model of an egg painted to match the natural colours of its own eggs, makes frantic attempts to sit on it rather than on its own eggs.

Theoretically, releasing stimuli will not trigger innate behaviours until the bird is motivated—*i.e.*, has built up a specific action potential through hormonal actions and other factors to perform behaviours. This build-up toward discharge of most any innate behaviour involves restlessness and exploratory maneuvers referred to as appetitive activity. Once the bird is sufficiently motivated, it then responds to the approprirte releasing stimuli with appetitive behaviour. Provided the bird's searches are successful and the stimuli act on its internal releasing mechanisms, the build-up of action potential acquired by the bird is then discharged in the performance of the consummatory act which consumes the energy.

Two illustrations of appetitive behaviour leading to the consummatory act are in feeding and nest-building. As the motivation (drive) to feed or to nest develops, the bird shows appetitive activity by searching for food or for nest site and nesting materials. When, finally, by responding to stimuli, it finds and swallows the food, or it selects the nest site and builds the nest, it performs the consummatory act. For the immediate future the bird has no desire to eat or to build another nest.

Occasionally, a bird performs an act in the absence of a discernible stimulus. This suggests some sort of build-up of motivation to the point where, in the continued absence of a proper stimulus, the energy is discharged in a behaviour called a vacuum activity. A good example is a male passerine bird bringing food to the eggs in his nest, "anticipating" the feeding of young. If the incubating female is absent at the time, he makes futile attempts to offer the food to the eggs and then departs, swallowing the food beforehand or taking it with him.

Often a certain behaviour appears irrelevant, being "out of context." An incubating bird, disturbed and driven from its rest by an intruder, does not attack, perhaps from fear, but instead proceeds to feed when it is actuallly not hungry or to preen when its plumage is in no need of attention. Two shore birds suddenly stoop fighting and one of them tucks its bill under the scapulars as though it were going to sleep. These are examples of what is commonly called a displacement activity.

Now and then a bird directs a certain behaviour at an object to which it does not normally respond. Two Herring Gulls, in a territorial dispute, vent their hostility by pulling grass instead of fighting each other. An incubating bird, disturbed at its nest by a human being, may attack another bird in the vicinity when it would not ordinarily notice that bird, or it may peck vigorously at the branch of a tree as if in anger. Any such behaviour is called a redirected activity.

Rather frequently a bird starts to perform a behaviour but does not complete it. For example, a bird crouches as though intending to spring into flight, but it does not fly. A behaviour of this sort, very low in intensity, is called an intention movement.

These behaviours—displacement activities, redirected activities, and intention movements which so often appear in conflict situations—are of special significance in that many seem to have become incorporated through evolution in more elaborate behaviours such as mating displays. They are then highly stereotyped or "formal" and are commonly referred to as ritualized. In this new role they serve as signals to elicit a response is another bird.

When any one of these conflict behaviours becomes ritualized for mating, the action is more exaggerated, more emphatic, and is consequently more conspicuous. Frequently, the conspicuousness is enhanced by bright colours in the plumage involved. Displacement preening in ducks such as the Mallard (Anas platyrhynchos) consists of lifting sharply and spreading the secondaries and adjacent scapulars of one wing and, while turning the head under the secondaries and scapulars as if intending to preen, revealing the iridescent speculum. Ritualized behaviour, in addition to being more conspicuous, tends to be species-specific—*i.e.*, peculiar to the species performing it. Thus displacement preening in the Mallard differs in minor ways from that in other ducks, thereby serving as an isolating mechanism that prevents cross-breeding.

Learned behaviour is essentially adaptive behaviour resulting from practice or experience.

Basic to learned behaviour is, of course, inherited behaviour which often predetermines the extent to which learned behaviour may develop. Learned behaviour, however, is frequently so fused with, and inseparable from, inherited behaviour that most behaviours cannot be declared the result of inheritance or learning or both. Thorpe provided a good example from his research on the development of song in the Chaffinch (Fringilla coelebs). He discovered that Chaffinches reared in auditory isolation could give extremely simple songs (subsongs), not the full song of the species. The simple songs were obviously inherited, but why not the full song? He found the answer when he exposed isolated Chaffinches at the age of about six months to recordings of the full Chaffinch song. They could learn to sing it. Yet other Chaffinches which he reared in isolation and exposed, at the same age, to recordings of songs of different species could not learn them well, if at all. Thorpe concluded that Chaffinches must learn the full song from their own species. It is only the components of the simple songs—in a sense, the ability to sing their own songs—that they inherit. They are born with no appreciable ability to learn the songs of other species.

The study of learning in birds poses numerous problems in interpretation and identification. Certain behaviours are sometimes misconstrued as the product of learning when learning is actually not involved. This is the case with mainly neuromuscular activities (motor activities) such as flying. A young bird just out of nest does not fly as expertly as one a little order, presumably because it has to learn how to fly. Yet a pigeon raised in a tube, without being able to use its wings, to the age when it should fly expertly, with indeed fly expertly when first released. Flying ability is inherent; it shows up not with practice or experience but with the maturation of neuromuscular coordinations.

FIG. 5.1: A territorial dispute between herring gulls.

Learned behaviours resulting from practice or experience usually develop in the early stages of a bird's life. Thorpe (1961) found that young Chaffinches must hear the full adult song of the species during their first year of development, if they are to learn it and sing it normally thereafter. Not hearing it, they may never learn to sing it even though they may be exposed to it later. A learned response in birds sometimes shows up long after the process of learning has taken place. Thus Chaffinches exposed to normal full songs as wild juveniles, then captured in the fall of their first year and isolated singly so that they cannot hear any other Chaffinch sing, may not show the effect of the original exposure until they begin to sing at the start of the first breeding season.

Learning takes different forms, all of which are attributed to experience, not maturation, and all of which may intergrade.

Habituation, the declining in a response to repeated stimulation without reinforcement ("reward or punishment"), is one form of learning. Stated another way, habituation is learning to avoid commonplace stimuli that are irrelevant to the welfare of the individual. Stated still another way, habituation is becoming used to, and no longer reacting to, anything new that proves to be harmless or non-rewarding. Tameness in a bird is the result of habituation whereby the bird becomes accustomed to unnatural conditions such as a cage or the heretofore unfamiliar actions of human beings. Once the bird is tame it is likely to remain so.

Another form of learning is trial and error, the selective acquisition during appetitive behaviour of a response resulting from reinforcement. A bird learns to swallow palatable objects and to reject objects with a bitter taste. In the same way, a bird when first flying learns to take off and land in the wind rather than with it and to select perches that can accommodate its feet and support its weight; a bird when first building a nest learns to select those materials that it can manipulate best. Playing (see below under Individual Behaviour) in a young bird may be a means of learning through trial and error.

A third form of learning is imprinting which usually takes place during a brief, critical period in the developmental stage of a bird's social life. Once acquired, the learned habit may sometimes remain stable for the duration of its life. Occasionally the behaviour manifests itself or reappears long after the process of learning. Imprinting is commonly observed in waterfowl and gallinaceous birds. Ducklings and goslings, for example, will learn to follow a substitute "parent" such as a human being, thereafter ignoring its natural parents. In the Mallard the period of "learning to follow" is between 13 and 16 hours. With age, or with the tendency to flee strange objects, the following response is apt to weaken, even disappear, though it may show up again under favourable circumstances. Imprinting is much less commonly observed in passerine and other birds reared in nests. Nevertheless, Thorpe has suggested a form of learning, probably imprinting, in Chaffinches. The young Chaffinch, hand-reared in isolation, sings simple songs that are innate; but the young Chaffinch, reared in the wild, on hearing the songs of its parents or other Chaffinches, acquires, in due course of time, characteristic features of the full song without apparently practising them. Later, in its first breeding season, it learns further characteristics from competing males. After 13 months, however, it can no longer learn features of other songs because its period of learning—presumably its period of imprinting—has ceased.

There are a few species of birds which manipulate or actually use objects to solve simple problems. The Tailorbird (Orthotomus sutorius) of southeast Asia forms a receptacle for its cocoon-like nest by perforating the edges of broad leaves with its bill and then drawing them together with plant fibres. A number of passerine species have the ability to pull up string to which food is attached. From the air the Blackbreasted Buzzard (Hamirostra melanosternon) of Australia drops stones onto emu eggs in order to break them open, and from a standing position the Egyptian Vulture (Neophron percnopterus) tosses stones onto ostrich eggs for the same purpose. The Woodpecker Finch (Cactospiza pallida) of the Galapagos Islands has the singular habit of picking up or breaking off cactus spines or twigs and using them as tools for extracting insects from holes and narrow crevices. Whether or not the bird acquires this habit by learning is open to question. Millikan and Bowman suggest that the habit may be a form of displacement activity "when another behaviour performance has been thwarted in some way." Possibly the ability is simply an inherent behaviour re-adapted through experience to perform a task more efficiently. Only a careful study of the way the habit develops among several individuals can prove whether or not learning is involved.

A few experiments have shown that birds possess quite remarkable abilities to learn, presumably by insight—the "sudden adaptive re-organization of experience." For example, Pastore proved with a series of 21 experiments on Canaries (Serinus canarius) that they can learn to respond to a unique object among a group of nine, the other eight being the same in all respects. For his experiments he put before each bird a board with nine small depressions in a row. In the first trial of his first experiment he capped eight of the nine depressions with aspirin tablets and the other with the head of a wood screw, the unique object, under which he hid some food, the reward. In the next trial he hid the food under an aspirin tablet, this time the unique object. In later trials the food was placed under either an aspirin tablet or a wood screw in depressions selected at random. The purpose of all the trials was to train the birds to find the food by pushing off the unique object. Each of the four Canaries, after an average of nearly 160 trials, achieved the ability to make the right choice in 15 out of 20 attempts. In a subsequent series of 20 experiments, using differing pairs of objects, the average number of trials required by the Canaries to choose the unique object became steadily fewer, to under 30 by the 13th in the series of experiments. Obviously, the Canaries learned to respond to uniqueness although they must have perceived it as abstract quality.

As another example of what may be learning by insight, Kohler taught a Common Raven (Corvus corax) to "count" an "unnamed number" of objects. To do this he trained the bird to open the one box in five which had the same number of black spots on its lid as on the key card on the ground in front of the box. The raven eventually learned to distinguish between five groups of spots—two, three, four, five and six. To prove that the bird was choosing the lid with the number of spots corresponding to the number on the card, Kohler changed the size, shape, and arrangement of the spots on both the lids and the card in a random manner. Even the position of the boxes in relation to the key card was changed. With this elimination of extraneous cues, the raven could still distinguish numbers of objects, but without consciously recognizing the numbers of such.

INDIVIDUAL BEHAVIOUR

Many behaviours are directed toward the bird's own welfare. These include maintenance behaviours, concerned with the care and comfort of the body, and behaviours associated with selection of habitat and cover, with food and feeding, and with playing.

Maintenance Behaviours

A large number of movements relate to the care of the plumage and, to a lesser extent, the skin and soft parts. Since the feathers, especially those used in flight or as insulation for the body, are vital to the bird's survival, the movements directed toward their care have evolved through natural selection to an elaborate degree and become highly stereotyped. Some of the maintenance behaviours have served as fruitful sources for displacement activities and displays.

Preening

Probably the most important as well as the most frequent activity in plumage carge; performed by the bill. Usually a bird deals with one feather at a time, seizing it at the base between the mandibles and working toward the tip, either nibbling it continuously or drawing it through the mandibles in one movement. This serves to clean the feather, to interlock barbs that have become separated, and to smooth down the plumage. Less often the bird moves the bill rapidly along the feather from tip to base near the shaft either in a stroking action or in a trembling action. When preening, the bird fluffs the feathers to make them more accessible and turns its head toward the plumage areas needing attention. Meanwhile, the bird contorts inself and assumes numerous postures, all quite stereotyped.

Head-scratching

Performed by the foot. Sometimes, along with preening, head-scratching is an activity for the care of the plumage, especially the feathers of the head and upper neck that are inaccessible to the bill; otherwise it is a simple reflex action in response to an irritation in the head region and is therefore a comfort movement. Birds use two methods in reaching the head: directly, by lifting the foot straight up from under the wing; indirectly by bringing the foot up and over the lowered wing. As a rule, different taxonomic groups of birds use on method or the other. Neither is reliable as a taxonomic character since some closely related groups, even some individuals of the same species, use both methods.

Bathing

This activity differs in water birds and land birds. Water birds (*e.g.*, loons, grebes, cormorants, and waterfowl), while floating, immerse their heads and shoulders in a scooping motion which sends the water over their backs while they rub their sides and flanks with their heads, ruffle their feathers, and lift and splash more water over their plumage by beating their wings. Wading birds stand in the water but bathe the same way. Land birds, going into shallow water, first dip their throats and breasts and flap their wings up and down, then raise their heads and fore-parts and lower their tails and hindparts into the water and flap their wings in and out of the water and over the back, spraying their plumage which is ruffled so that water can reach all the feathers and even the skin. There are many exceptions to this method of bathing. Swifts, swallows; and other highly aerial birds, as well as hummingbirds, kingfishers, and some owls, drop repeatedly to the water's surface and perform a skimming action sufficient to wet their plumage. Swifts bathe in falling rain. Tropical species of many varieties bathe mainly in the rain or wet foliage. Almost all land birds will take advantage of rain or even lawn sprays. Bathing functions as a means of keeping the plumage in order, rarely, for cleaning it. While bathing, a bird is liable to predation; hence it is innately apprehensive and habitually avoids drenching itself to the extent of impeding its escape by flight.

Oiling

A form of preening among birds with an oil gland. Typically and often in association with preening, the bird touches or pinches the oil gland with its bill, stimulating the flow of oil to the bill's surface. The bird then smears the oil on the plumage areas with rubbing movements of the bill. But there are many variations. Long-necked birds often first rub the head of on the oil gland and then rub it immediately on the plumage areas. Short-necked birds, such as passerines, oil the head with the foot, first taking the oil with the bill, then scratching the bill with the foot.

Sunning

An activity among practically all birds, sometimes performed after preening, bathing, or dusting. Typically, in a moderate reaction to warm sunlight, passerine bird raises the feathers of the head, neck, back and rump, crouches with the axis of the body perpendicular to the source of light, spreads the flight feathers of the wing nearer the sun, and fans the tail. It sometimes pants with opened mouth. It may draw the nictitating membrane over the eye and perhaps close the lower lid. Non-passerine birds react in different ways. Vultures and other large diurnal birds of prey may stand, facing the sun, with wings extended. Cormorants sun themselves similarly after coming out of the water or long after their plumage is dry. Birds being photographed and exposed too long under bright lamps that radiate considerable heat, give the normal sunning response. During moderate to high-intensity sunning, some birds seem to be in a semihypnotic state and much less cautious than is normal. Whether sunning directly benefits the plumage or serves any other significant function is undetermined.

Dusting

A behaviour in which dust—powdery dry surface soil—is forced through the plumage. In an average performance, the bird proceeds to scratch and/or peck out a depressicn while crouching and usually turning its body. Once the hollow is deep enough and the soil loosened, the bird sits in the hollow, raises dust through the ruffled plumage with a flicking and scooping action of the wings alternating with a scratching action of one foot or the other. Intermittently, the bird compresses its feathers and pauses to peck dust against itself and under the body and, at least in the gallinaceous bird, to lower and rub its head and neck in the dust. On ceasing to dust, the bird stands, ruffles its plumage and vigorously shakes out all dust particles. Unlike bathing, dusting is characteristic of comparatively few birds. It appears to have developed commonly among birds of deserts and grasslands. Some birds—*e.g.*, wrens, House Sparrows (Passer domesticus)—bathe as well as dust; others such as gallinaceous birds and goatsuckers dust only, rarely if ever taking a bath by deliberately standing in water. Dusting no doubt removes ectoparasites and may have other functions as yet undertermined.

Anting

A behaviour in which the fluids from ants, notably formic acid, are applied to the plumage and possibly the skin. There are two basic types of the behaviour: active anting, in which the bird anoints its plumage with the bill, and passive anting, in which the bird allows ants to invade and anoint the plumage. Whitaker has reviewed the subject and described active anting in the Orchard Oriole (Icterus spurius). In a typical performance, the oriole crushed the ants and dabbed them, with rapidly vibrating head, to the ventral surface of the wing tips and especially the under tail coverts and bases of the rectrices. It anointed the feathers of the belly and tibiae only occasionally. A regular feature of the behaviour was tripping and tumbling, the result of bringing the tail forward beneath the body. In passive anting, as described by Simmons, "birds squat or lie down among the ants, often while applying them directly and going through the motions of doing so with empty bill, while the wings and tail are partly or wholly spread out." Whitaker listed 148 species, 16 of them non-passerine, which have been captivity, perform active anting with burning matches and cigarette smoke, moth balls, hair tonic, mustard, citrus fruits, numerous insects other than ants, and other substitutes. What causes anting and what purposes or purposes it serves, if any, need investigation. V.B. Dubinin discovered that the formic acid from ants may destroy feather mites and thus he believes that anting functions as a defense against them.

Comfort Movements

Any number of actions which presumably give the bird "comfort." Some common movements are feather settling by first raising the feathers, shaking the body, flapping the wings, and then depressing the feathers into proper pasition; stretching, often by extending the leg and wing of one side while fanning the tail on that side, then repeating the action on the other side, yawning, opening and closing the free-moving mandible, in most birds the lower one; and resting, by standing on one or both feet, or by squatting down, the plumage relaxed, the head retracted and seeming to rest on the shoulders, sometimes turned with the bill tucked under the scapulars in a sleeping posture.

Sleeping

Birds ordinarily sleep with the head retracted and turned so that it rests on the back with the bill tucked under the scapulars. But there is both intraspecific and interspecific variation. Some individuals may sleep with the head retracted to the shoulders on which it seems to rest and the bill forward.

Several familial groups, such as pigeons, normally sleep this way.

If not sleeping when floating on water, birds usually squat rather than stand. If squatting on a perch, the skeletal and muscular arrangement of the hindlimbs is such that the toes automatically grasp and hold to a twig or branch. Goat-suckers though lacking this arrangement, can sleep in trees by squatting with the body lengthwise on a wide branch.

Instead of squatting during sleep, a few groups of birds such as woodpeckers, swifts, and presumably creepers (Certhiidae) cling to vertical surfaces and a few others such as the mousebirds of Africa (Coliidae) and some parrots (Loriculus) hang upside down from perches.

Selecting Habitat and Shelter

Most species of birds select distinct habitats. This allows an efficient use of natural environment and assists in reducing competition between species. Birds no doubt recognize their habitats by visible cues, but what are the cues and how do preferences become established in the first place? To answer the question, one may identify the components of an environment and then determine how the preference develops by comparing individual birds raised in captivity under different conditions with wild trapped birds. Klopfer did just that with Chipping Sparrows (Spizella passerina), taken from fixed stands of pine and oak in the North Carolina Piedmont and kept in a chamber where they were exposed to both fresh pine and oak foliage. Since Klopfer designed the study to measure the bird's preference of foliage for perch sites, all other environmental variables except foliage and light were minimized as much as possible. Klopfer found that adults, trapped in the wild, and young, reared in isolation with no foliage, both preferred pine over oak; but the young, reared in isolation with oak leaves, preferred oak over pine. Their choices suggest that size and shape of leaves and to a less extent light-shadow patterns are the cues and that the preference of one foliage over the other, though innate, may be modified by experience.

Klopfer's study does not suggest that the selection of a habitat constitutes an innate response only to such simple features as size and shape of foliage. Almost certainly the selection depends on many cues, single or in combination. However, Clopfer's work does quite properly imply that birds, while innately preferring a particular habitat, are nonetheless genetically equipped to re-adapt to others and thus demonstrate considerable plasticity or oppotunism. Field observations support this. For example, in northern Michigan the Wood Thrush (Hylocichla mustelina) prefers cedar-spruce bogs to the deep deciduous woods one would expect. In West Virginia the Magnolia Warbler (Dendroica magnolia) appears in oak forests whereas northward in North America it is ordinarily partial to spruce and hemlock forests.

Klopfer's study as well as field observations tend to support the concept that birds breeding in the temperate areas of North America show opportunism because their environment is unstable and they must cope with variability. Birds breeding in tropical areas are less opportunistic because their environment is comparatively stable.

The habitats which birds select must meet requirements of shelter favourable to their survival, day and night, in all seasons of their occupancy. Besides shelter that provides escape and protection from predators, there must be shelter for roosting. Sometimes one shelter may serve two or more purposes—*e.g.*, a shelter for escape may also be a nesting shelter or even a roosting shelter. But whether serving one or more than one purpose, birds are nevertheless as selective of these shelters as they are of the overall habitat.

A few remarks about roosting shelter are appropriate since most works on ornithology give relatively little attention to the subject.

Klimstra and Ziccardi found that covies of Bobwhites (Colinus virginianus) in southern Illinois much preferred to roost on well-drained ground at medium to low elevations facing south or southwesr and protected by a low, sparse, canopy of vegetation. Any shift in roosting ground was mostly from lower to higher elevations, except in windly and inclement weather when the shift was to poorly drained sites at lower elevation. Some cavity-nesting birds may use their own nests as roosts in the winter, or construct or excavate special nests for roosting. Berger watched a male Downy Woodpecker (Dendrocopos pubescens) in October excative a roosting cavity in a dead naple, much as it would a nesting cavity, and use it regularly through the winter until late April. Frazier and Nolan noted five to 14 Eastern Bluebirds (Sialia sialis) roosting together in a birdbox during the winter in Indiana. Many other observers have reported instances of two or more birds roosing together in such enclosures or even natural cavities. After a heavy fall of snow the Ruffed Grouse (Bonasa umbellus) will plunge into the snow and pass the night.

The above illustrations bring out several important points. Besides providing protection from predators, the roosting shelter gives protection from unfavourable weather and helps conserve body heat. When two or more birds choose to roost together in a small cavity, they naturally conserve heat. Also, in selecting roosting shelter, birds show opportunism as, for example, when choosing a nestbox instead of a natural cavity or when taking advantage of a snowfall.

Selecting Food and Feeding

Most species of birds have the structure and motor coordinations for taking a wide variety of food although they usually learn to select particular items. Among closely allied species in the same habitat this reduces competition and assures each species its share of the total food supply.

Each species is innately equipped with fixed action patterns for seeking food and preparing it for consumption. In the young bird these movements may be elicited initially by somewhat generalized stimuli. Gallinaceous chicks, for example, begin seeking food by pecking at small objects with colour values contrasting with the background. To test the food-pecking responses of young Indian Peafowl (Pavo cristatus) to objects of different sizes and colours against different background colours, Dilger and Wallen exposed four newly hatched chicks to circular objects ranging in size from two to 14 millimeters, painted in six different colours, and placed against backgrounds with the same six colours. The chicks preferred the smallest objects when given a choice of all sizes, presented in different colours in random combination with differently colours background. They also preferred the combinations of objects and backgrounds that offered medium rather than maximum or minimum contrast in both hue and brightness. In all probability the chicks were innately adapted to the medium rather than the maximum contrast that one might expect, because edible objects in the wild are more usually in medium contrast with the background. Dilger and Wallen suggest that maximum contrasts may be "somewhat inhibiting or possibly even somewhat frightening."

The size of the bill and its special adaptations are among the important factors in determining the food items a bird learn to select. In general, young birds select smaller items than older birds; as their bills harden and their motor coordinations mature, they take larger items though not beyond the point where they can handle them efficiently. Provided they have the necessary behavioural and structural equipment, which in many species included the feet, birds eventually learn from their own trial-and-error responses to generalized stimuli, or from watching their parents and conspecific associates, to select particular items in various local situations.

In England since 1911, titmice and other small birds—at least 11 species altogether—have learned the habit of drinking from milk bottles by either picturing metal foil caps, or stripping off or removing paper or cardboard caps. Some observers reported titmice in groups following milkcarts and attacking bottles immediately upon delivery. While many titmice no doubt learned the habit from other titmice, it is just as likely that it developed independently among others.

Playing

Birds perform playful activities though not to the extent that mammals do. Young birds still in the nest show exploratory pecking—gently pecking one another and the nesting materials. Chicks in a brood, after leaving the nest, similarly peck one another and peck at subjects on the ground. Occasionally, they pick up morsels of food and run with them, eating them later. Gull chicks when scarcely half-grown pick up, run about with, and drop sticks, plant stalks and other objects. All young birds, well in advance of flying, fan their wings from time to time. Such activities constitute what amounts to "practising" and may be a form of learning to forage, escape, of fly.

Fledglings and chicks of land species commonly make short flights or runs, turning sharply—an activity called playfleeing. Aquatic chicks give a comparable performance on the water by running, splashing and turning on the surface, even attempting to dive.

Adults of some species perform activities that have all the appearances of play and seem to serve no function other than the release of pent-up energy. At the Falkland Island in the South Atlantic, Pettingill observed groups of Magellanic Penguins (Spheniscus magellnicus) running from a beach into the sea, quickly submerging, torpedoing away from shore in a wide circle, and as quickly emerging and hustling up the beach. Parrots are notoriously playful. The Kea (Nestor notabilis), a large powerful mountain species in New Zealand, pulls windshield wipers off cars and tosses empty tin cans about for no appartent reason than to be doing something.

SOCIAL BEHAVIOUR

Social behaviour involves the many kinds of interactions or joint activities that have to do with two or more individuals in the same species or even individuals in different species.

Agonistic Behaviour

One of the most common forms of social behaviour is agonistic behaviour which includes all manner of hostile activities or displays from over attack to overt escape.

Perhaps most obvious hostile displays are those that have become ritualized as social releasers. The most abundant of these are probably the threat displays, "designed" to produce a "frightening" effect, forcing an opponent to retreat. Threat displays are thought to be activated by drives to attack and escape. For example, a bird approaching the territorial boundary of another exhibits attack, but, as it crosses the boundary, the motivation to escape sets in, coming into balance with the motivation to attack. Meanwhile, the owner of the territory, instead of being intimidated, proceeds to attack the intruder. If, at this point, the attack drive of the intruder still continuous more intense than its escape drive, an encounter or fight between the two birds results; if the escape drive of the intruder becomes strong enough to inhibit or outbalance its attack drive, the bird flees.

Threat displays are believed to be derived from a number of different sources, among them the unritualized intention movements associated with locomotion. In certain passerine birds the threat displays may include lowering the body to the horizontal, facing and extending the head toward the opponent, sleeking the head feathers, holding the wings stiffly down or perhaps flitting them, depressing and fanning the tail or sometimes flipping it. All such actions suggest the movements of a bird about to take flight.

Almost as common as threat displays are the appeasement displays serving to prevent attack without provoking escape. These displays serve to reduce the strength of the opponent's drive to attack and, to a lesser extent, the opponent's drive to escape.

As in threat displays, most of the movements have been derived from unritualized hostile sources, in this case, mostly from intention movements involving escape. These include, among certain passerine birds, sitting, flexing the legs, "hunching" with the head held low and retracted to the shoulders, and turning away from the opponent.

Vocalizations frequently complement both threat and appeasement displays, giving them an auditory component. Most songs of male birds function in part as threat displays, warning and/or repelling rival males.

Threat displays serve to space out birds, thereby preventing over-population in any one area. Appeasement displays, on the other hand, birds closer together by inhibiting attack or aggression and promoting bonds between pairs for reproduction and between parents and young.

Defense Behaviour

All bird species are subject to predation and, consequently, show behavioural adaptations that promote their survival. Like maintenance behaviours, the defense behaviour of the individuals directed toward itself, in this case its own safety, but it may also serve to promote the survival of other individuals—*e.g.*, the individual's own young, also the individuals of its own species and/or other species. Thus defense behaviour, under given circumstances, can be a form of social behaviour.

Birds respond to certain danger stimuli both auditory and visual. Unaccustomed sharp sounds and abrupt movements evoke defensive behaviour of some sort. Models of owls that have the right combination of shape, colour pattern, and contour can elicit alarm reactions in small birds. Cardboard models, cut in the shape of a short-necked, avian predator flying overhead and passed over the birds in the direction of the short neck, can induce escape reactions in waterfowl. The exact shape of the wings and tail and the colour are immaterial. Although it was once though the responses to models were simply innate, more recent researches have demonstrated that at least some learning is involved.

The precise manner in which birds respond to danger signals depends on the physical and behavioural adaptations of the species responding and the type of predator, the intensity of signals, whether or not the responding individual is breeding, and the circumstances of the encounter. Some of the more common responses are the following:

Fleeing and Freezing

If the average passerine bird is exposed without benefit of cover and the danger signal, such as a hawk in flight, is strong, the bird flees to the nearest cover where it is likely to "freeze" —*i.e.*, to keep completely motionless with the feathers sleek ed, the legs flexed, the head down in line with the body but cocked so that one eye is fixed on the direction of the predator. If the bird is already in cover when receiving the danger signal, it freezes where it is. The freezing posture is concealing and, at the same time, practically identical to take-off posture; thus the bird is prepared to flee if necessary.

Birds such as grouse, snipe, woodcock, and goatsuckers with cryptic colouration respond to a strong danger signal by freezing even when not in a shelter. Most chicks, nestlings, and fledglings of all species normally freeze in response to a danger stimulus and often close their eyes.

Communicating Danger

The vocal repertoires of practically all birds include special alarm or warning calls, given in response to danger signals, but not necessarily in all cases or at all times. A passerine bird at the sight of a cat invariably utters alarm calls, yet the same bird may freeze at the sight of a flying hawk and remain silent or give alarm calls only if it has young. Though presumably not consciously given to alarm or warn, the calls nonetheless alert flock associates and mates and induce freezing among young.

Threatening

Confronted with potential predators or animal forms that impose danger, birds may respond with actions that threaten or intimidate. In some instances the actions may appear identical to the threat displays against members of their own species, but often they are more exaggerated or more elaborate. Threatening behaviour includes actions that make the performer seem more formidable. Thus most birds puff up their feathers, appearing larger than they are; and a great many birds open their mouths as if threatening to bite, and lift their wings, as if intending to strike. Some birds utter infrequent sounds such as hisses that may function as an element of surprise and deter an approach. A Great Horned Owl (Bubo virginianus) not only hisses but also snaps its bill rapidly; a Least Bittern (Ixobrychus exilis) silently lifts its bill skyward and sways back and forth.

Birds with young most often employ threatening behaviour and may vary the type and intensity with the kind of predator. A bird that performs vigorously at the approach of a small mammal may flee at the approach of a hawk or a human being.

Attacking

Many birds with nests or young, instead of threatening a predator, may attack it outright. If threatening behaviour fails to intimidate a predator, attacking behaviour may follow. In a sense, attacking behaviour is a projection of threatening behaviour.

Exploring

Occasionally, birds show a tendency to explore strange objects, such as a blind suddenly placed near their nest. While examining it visually from different angles, they exhibit what appears to be a conflict between the urge to fly away (escape) and the urge to see more. Exploratory behaviour of this sort may have survival value by leading to the discovery of incipient danger.

Mobbing

In any season, groups of small birds respond to a perched hawk or owl by "mobbing" it. During the breeding season, birds often leave their territories temporarily to join in the performance. The basic motivation may be combination of two tendencies: to explore and to escape. In some instaces, the tendency to escape may be supplanted by the tendency to attack outright. The function of mobbing is puzzling. Possibly it serves to put all birds in a given area on the alert and, during the breeding season, to distract a predator from discovering young birds in the vicinity.

Giving Distraction Displays

Parent birds, threatened by a predator, may perform movements that sometimes simulate incapacity from injury ("injure-feigning") or sometimes suggest the running away of a small mammal ("rodent-running"). The motivation is believed to be a combination of two crives—to escape and to attack—that was become ritualized and functions in diverting the attention of the predator from the nest or young to the parents.

Injury-feigning shows up in a great number of species representing many families. The principal movements consist of fanning, beating, or dragging one or both wings, spreading and depressing the tail, fluffing the back and rump feathers, giving "distress" calls, and alternately moving parallel with and away from the predator. The performance differs in detail from species to species depending to some extent on the habitat. In general, species nesting on the ground most fully perform injury-feigning, but species nasting in other situations, such as trees and marshes, at least use some of the movements and adapt them accordingly.

Rodent-running is characteristic of species nesting on the ground in thick grassy cover. Essentially it amounts to the bird sneaking off the nest, away from the predator, and running fast, occasionally hopping slightly and even fluttering its wings as it disappears from sight.

Full distraction displays at maximum intensity are given when the eggs are hatching and for one or two days thereafter. While some predators may elicit distraction displays in a species, others may instead elicit outright attack. If a displaying bird is harried by a predator or a human being for an extensive period, its performance gradually wanes in intensity and may cease altogether.

Flocking Behaviour

The majority of birds tend to be gregarious at one time or another. During the breeding season, most sea birds (including pelicans, cormorants, gulls and terns) and certain species of herons, swallows, and tropical icterids form colonies in which pairs establish and maintain nesting territories close to one another. Outside the breeding season a great many birds gather in flocks in which they remain until they disperse to nest. A few birds, such as the Monk Parakeet (myiopsitta monachus) of Argentina and certain species of weaver finches (Ploceidae) of Africa, are gregarious at all times, nesting in colonies and staying in flocks when not breeding.

Flocks may be composed of one species or several species of similar size with approximately the same flight speeds but not necessarily the same food preferences and feeding adaptations. If a flock is of one species, as is more often the case, it may be comprised of one family or multiples of families from the preceding breeding season. This is believed true with small flocks of migrating geese. Indeed, at Delta, Manitoba, coloor-banded adult and young of the year Canda Geese (Branta canadensis) were seen "to depart southward together in Novermber and return, still as a united group, the next April." This same make-up may also be true of small flocks of

FIG. 5.2: Predator Mobbing, A noisy flock of crow mobs a golden eagle. The mobbing crows are at little risk of being captured by the eagle, but the eagle findes it difficult to concentrate on its hunting and is likely to move elsewhere. Birds normally mob only those predators who are able to launch a surprise attack. Mobbing calls the attention of most individuals in the vicinity to the presence of the predator, so none of them are likely to be taken by surprise.

permanent-resident chickadees (Parus spp.) during the winter. But there are numerous exceptions. For example, among shore birds migration southward from the Far North, adult birds constitute the first flocks; birds-of-the-year come in later flocks. Early spring flocks of Red-winged Blackbirds (Agelaius phoeniceus) are likely to be made up entirely of males.

Each bird in a flock very rarely, except when roosting for the night, comes in physical contact with its fellows, but maintains its individual distance. Cliff Swallows (Petrochelidon pyrrhonota), perched on a wire, keep about four inches apart. Any individual attempting to sit closer is immediately threatened by the bird it is crowding.

Numerous signals such as peculiar vocalizations and colour marks maintain the integrity of a flock. To these signals each individual responds. Should an individual become separated from the flock, it quickly shows appetitive behaviour by calling and searching until it finds its flock members.

Within a closely integrated flock all the activities—feeding, preening, bathing, etc.—and movements tend to occur simultaneously. Usually one individual initiates the activities and the others soon perform them too—a procedure called social facilitation because the performance of one individual stimulates the behaviour of its fellows to act similarly. Through social facilitation a bird will sometimes eat more than its normal capacity by seeing one or more of its associates feeding. The concerted, almost instantaneous movements in a flock are attributed to a "following" reaction, the instinctive urge among gregarious birds to stay close together. There is no better example of this trait in operation than in a compact group of shore birds suddenly rising as in a body from a mudflat, wheeling over the area in perfect unison, and setting again, still compact.

In small flocks of conspecific birds, whether caged or in the wild, a social arrangement may be established that is normally hierarchial with each individual having its own rank in descending order of dominance. Schjelderup-Ebbe first brought this phenomenon to light in the Domestic Fowl (Gallus gallus). Studying penned birds, he discovered the existence of a peck order wherein Bird A can peck all the others and is therefore dominant, Bird B can peck all but Bird A, Bird C can peck all but Birds A and B, and so no until there is one bird that can peck no other and is subordinate to all. Instead of a direct-line peck order. there can be a triangular or polygon arrangement in which Bird A pecks B, B pecks C, and C pecks A. This situation is apt to be only temporary and may revert back to the direct-line hierarchy.

The position, rank, or peck right of the bird in the hierarchy is determined by one or a combination of factors such as sex, vigour, stage of molt, colour, and health. If something happens to the bird—for example, if it is injured and thus physically impaired—it may lose its peck right in the hierarchy and become the subordinate bird. Should the bird at the top of tlae hierarchy be removed, a scramble ensues in the flock for top position. A newcomer to a flock with an established hierarchy can work its way in social rank only by starting as the subordinate bird.

In some flocks of conspecific birds the peck order is less fixed. Instead of a bird having a peck right over another and making that bird an absolute subordinate, it may achieve a peck dominance over another by winning more encounters than it losses. This may be observed in flocks of chickadees or Tree Sparrows (Spizella erborea) at a feeding station. Among heterosexual flocks of birds there mny be more than one hierarchy. In a caged flock of 12 Red Crossbills (Loxia curvirosta), Tordoff noted three hierarchies, a peck order of males, a peck order of females. and a peck dominance of males over females. The lowest-ranking male was most active in dominating females, followed closely by the top-ranking male. In a wild flock of Ring-necked Pheasants (Phasianus colchicus), Collias and Taber found a dominance order among both sexes, where all the cocks dominated all the hens until the onset of the breeding season when they stopped pecking them.

Heterospecific flocks of birds show a species hierarchy geuerally established in accordance with the size and or hostili y of the species comprising it. Colquhoun, for instance, observed that, when flocks of nuthatches and titmice were feeding, the European Nuthatch (Sitta europaca) usually dominated the Great Tit (Parus major), the Great Tit dominated the Blue Tit (P. caeruleus), and the Blue Tit dominated the Marsh Tit (P. palustris).

Whether or not a flock has leadership remains a moost question. A family flock of Canada Geese is commonly believed to be led by the adult male, and a migrating wedge of the same species made up of multiple families is thought to be led by one of the older males, but all this is supposition. Hanson found that the adult male Canada Goose will often initiate a local flight of his family put when the family is on the ground or in the water, the fomale often leads the flock while the male guards it at the rear. If there is a leadership in most flocks of adult birds, there is no evidence that it is taken by the top-ranking birds in the hierarchy. Quite possibly any persisting leadership of a flock of fully mature birds is non-existent since, as Edmund Selous once suggested, the flock is an individual, acting, moving, and behaving as a "collective bird."

Flocking behaviour has survival value in at least two important respects:

1. B.rds close together with integrated behaviour are less vulnerable to predators than solitary individuals. With more eyes to watch, the flock has a greater awareness of the enemy's approach. When threatened, flock members acting in concert can take deierrent action. Either they can draw themselves together in a formidable mass as ducks resting on a pond converge to form a tight raff, or they can attack with a mobbing action as do gulls and terns in a nesting colony.

2. Birds foraging in a group increase the chances of success in finding and exploiting soucrces of food.

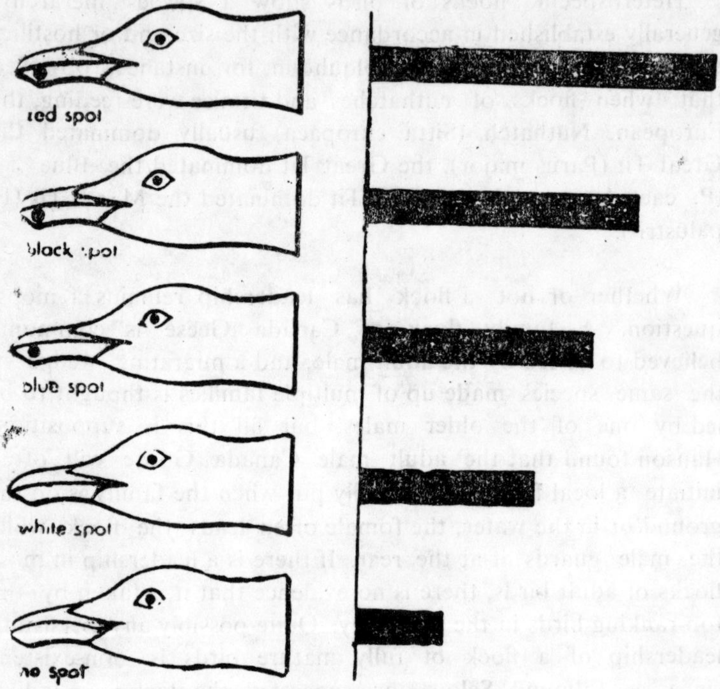

NUMBER OF RESPONSES

FIG. 5.3: Herring-gull nestlings tap at the red spot on the bill of the parent; the parent then feeds them N. Tinbergen presented cordboad models of the adult's head to newly hatched chicks and obtained the results indicated in the graph. Evidently, innate behaviour depends upon the nature of thestimulus.

Reproductive Behaviour

The interactions of birds that have to do with territory, mating, nesting, and parental care of the yonng come under the heading of reproductive behaviour, treated in later sections of this book.

GENERAL STUDY PROCEDURES

The science of behaviour, or ethology, is essentially the study of animal activities by direct observation, supplemented by experimental methods. While ethology is not exactly new, it does represent "return-to-nature" approach in understanding the function of the live animal.

Although much is being learned rapidly about the behaviour of birds, was gaps in knowledge remain, even of the most common species. Avian ethology is still a wide open field for investigation.

In studying the behaviour of any bird species, the student must describe and name its activities or behaviours. In other words, he must know thoroughly the behavioural make-up of his subject and be able to designate the different components.

Describing Behaviours

Describing a behaviour is much more difficult than describing form, shape, colour, and other morphological features. As Dilger stated, behaviour is ephemeral—it takes place and is gone. It may last a fraction of a second, or, like sleeping, it may last for hours. Moreover, a behaviour not only varies normally as does structure, but it also varies in the individual bird, depending on the subject's physiological state, motivation, and many other factors. An adequate description of a behaviour must, therefore, be based on many more samples than would be required for structure.

Compounding the difficulty in describing behaviour is the lack of suitable terms and the consequent recourse to words that describe human activities. This frequently results, however unintentionally, in ascribing to a bird certain human traits and motives. In any science treatment of behaviour always try to avoid anthropomorphisms—*i.e.*, statements that suggest human thoughts, attitudes, and reactions—and concepts that involve teleology—the doctrine that gives a conscious purpose or objective to any activity undertaken.

Also compounding the difficulty in describing a behaviour is its complexity, no matter how simple it may seem. Any behaviour may often be comprised of two or more separable acts, each of which is a set of observable activities, seen in combination and not analyzable into separately occurring components (Russell et al., 1954).

Dilger described a good example of a particular behaviour and its component acts in the Peach-faced Lovebird (Agapornis roseicollis) which carries strips of nesting material in the feathers of its lower back and rump. When the bird puts a strip of paper in the feathers, first it points the strip either to the right or to the left. Next, the bird tucks the strip into the feathers of the lower back by turning its head over one shoulder and bringing its bill with the strip into the feathers, which are simultaneously raised. Then the bird thrusts the strip hard into the feathers with a trembling motion of its head, releases the strip, and brings its head forward while compressing the feathers. In all, Dilger noted at least eight separate acts performed in a characteristic manner and sequence. At the outset of his observations he considered thrusting and trembling a single act but, with later experiments, he found them to be separate because some birds trembled without thrusting.

Naming Behaviours

The established custom is to derive names that serve to describe or at least indicate the most prominent action. If the behaviour is a display—*i.e.*, a behaviour that has evolved as a signal or releaser—the name is usually capitalized as, for example, Wings-up or Bill-up-and-open. But if the behaviour is of any other sort, the name is not capitalized—*e.g.*, bill-down or tremble-thrust. As a rule, it is inadvisable to include in the name of a behaviour any word which designates its function or factors responsible for it—*e.g.*, wings-up threat or aggressive bill-up-and-open. Such an inter-pretation is unwarranted, may eventually be proved wrong, and may bias future investigations.

Objectives of the Study

Having described the behaviours of a bird, the student may

proceed to postulate their function, survival value, causes, and origin.

Function

What a behaviour accomplishes. It may simple serve the individual performer, as in the case of a maintenance activity or in a flight from one place to another, or it may, in the case of a display, serve to elicit a response in another bird. One can usually postulate a function of a display by nothing the kind of response elicited from the bird before which the display is performed.

Survival Value

Any normal behaviour in an individual bird is presumed to enhance the welfare of its species, otherwise the behaviour would not have evolved. The survival value of biological significance of many behaviours—*e.g.*, flying to a food source, escaping from a predator, etc. – may be so obvious it needs no attention, or it may be obscure. In that case, numerous observations of the behaviour, performed repeatedly under varying circumstances in the wild and/or in captivity, may provide the clue or at least the basis for making a reasonable guess. Why the occasional singing at night be certain species that regularly sing during the day? Why the teetering and tail-pumping action among many different species? Questions of this sort present a challenge to the avian ethologist.

Causes

A bird normally performs a behaviour from both internal and external causes. Internal causes include drive, usually conditioned by experience or learning; external causes may be one or more sign stimuli arising from the environment. The student can often determine external causes by careful field observations. In the natural environment the situation in which a behaviour occurs is rarely the same from one occasion to the next. Thus it is possible for the student to establish the cause by observing and comparing a long series of occasions, in which the behaviour is performed, and noting the occasions, when the behaviour is not performed. In a sense, this a "natural experiment." For example, what causes small birds to be alarmed by hawks on some occasions, to ignore them on others? Through observations, the student will discover that small birds generally show alarm when the hawk is in flight, but not when it is perched. The movement of the bird in flight is the principal cause of the alarm.

Another way of studying the external cause of behaviours is to compare similar behaviour apparently acquired independently by unrelated species—*e.g.*, the habit developed by Cattle Egrets (Bubulcus ibis) and Brown-headed Cowbirds (Molothrus ater) of following large, grazing mammals in search of insects stirred up by their feet. There is always the chance that a comparative study of a habit common to very different birds will yield the clue to its common cause.

Establishing the internal causes of behaviour and the role of experience in performing it requires sophisticated procedures under laboratory conditions. One method, to determine the role of experience, is the so-called deprivation experiment in which two groups of birds are used, one experimental, the other a control. The two groups are identical except that one group has been deprived of experience. By comparing subsequent changes in the performance of the experimental group to the "base line" of behaviour of the control groups, one can determine the extent to which experience plays a part in the behaviour.

Origin

How behaviours have evolved is an important objective in ethological studies. After describing and observing all the behaviours in a species, the student may postulate that certain displays are derived from maintenance activities, others from displacement activities, intention movements, and so on. By comparing certain homologous behaviours in closely related bird species, the student may discern different stages in the evolution of these behaviours—information that may also shed light on the evolution of the species themselves.

THE STUDY OF BEHAVIOUR

To give the beginning student in ornithology an introductory experience in observing behaviour, the following two studies are outlined below. For other studies that may be undertaken, see Stokes (1968).

1. Behaviour in a gull colony.

Procedure

Prior to the study, read The Herring Gull's world by Tinbergen (1953) with sufficient thoroughness to be aware of the various posturings and calls, the capacities of the adult bird's sense organs, the non-reproductive behaviours, and the more conspicuous behavioural features of territory establishment, pair-formation, incubation, and parent-chick relationships. Be equipped with notebook, stop watch, and (optional) a small portable tape recorder.

Erect a blind in the colony where several nests with eggs and/or small chicks are in view through one opening.

6

Social Behaviour in Carnivorous Animals

About 37 million years ago, when primitive fissiped carnivores were evolving from their miascid predecessors, some of them forsook their way of life on land and went to sea to find food. This move must have worked out well, because many of the descendants of these pioneers—the seals, sea lions and walruses—are still making a living in the sea today. No doubt, descendants of the carnivores that remained on land fared well, too. Because land and sea are such different environments, we might expect the two animal groups to have diverged in many ways since their separation. Indeed, it should be interesting to compare representatives of the two groups because of what they have in common, an ancestor, and becouse of the obvious differences in the environments in which they live. In this chapter, some differences and similarities aquatic caruivora like between seals and their terrestrial counterparts with respect to gross morphology, feeding and predatory behaviour, reproductive behaviour, and various aspects of social life have been discussed.

The similarities between aquatic caruivores and land carnivores were obvious to laymen and early scientists.

Fifteenth and 16th century sailors described the seals they saw as marine carnivores and called them "sea lions," "sea bears," or "lobos de mar" (sea wolves). Linnaeus and several later taxonomists classified seals, sea lions, and walruses in the order Carnivora. However, as early as 1811, there was an attempt to place these animals in a separate order (Illiger, 1811). No-a-days, many investigators class these marine mammals in their own separate order, the Pinnipedia. However, classification of these animals is far from settled. Some maintain that the new order is inappropriate and that pinnipeds should be considered merely a suborder of the Carnivora. Others argue that the pinnipeds can be classified into two living families (walruses are placed in the same family as sea lions) under the super-family, Canoidea, in the order Carnivora.

The earliest pinnipeds probably entered the sea in one of the following areas: the Arctic Basin, the northwest coast of North America, or the Tethyan-Mediterranean area. It is not cleaı whether seals and sea lions (including the walrus) were already differentiated at this time. The oldest remains of pinnipedare from the Miocene era, a time when they were already well adopted for aquatic life and when seals and sea lions were already distinguishable. One opinion holds that all pinnipeds derived from canoid or dog-bear stock. An opposing view is that sea lions and the walrus derived from ursine stock, and seals derived from lutrine stock. The question is far from settled. Nevertheless, there is agreement that all pinnipeds are more closely related phylogenetically to members of the superfamily Canoidea of the order Carnivora the dogs, raccoons, bears, weasels, and the like—than to the feloid carnivores.

After their enterance in to the water, the distribution and diversity of pinnipeds were influenced by geomorphic and climatic barriers, distance between land falls, ocean currents, and water temperature. Davies hypothesizes that pinnipeds are, and always have been, tied to a cold-water environment. He argues very persuasively that the distribution and differentiation of present-day northern pinnipeds reflects periods of expansion and contraction of sea and glacial ice that occurred during the Pleistocene era. Scheffer points out that "as local shore lines and islands rose and fell and glacial barriers came and went, pinnipeds moved back and forth in order to maintain favourable breeding grounds along the edge of the sea." On land, many contemporary carnivores were being forced into extinction by periodic fires, floods, droughts, and dust storms. Scheffer suggests that the early pinnipeds moved out along the shores of continents, from island to island and later to polar ice fields. The rate of evolution was faster at the frontiers of the advancing lines where immigration was unidirectional.

> The primitive populations were small and the members did not wander far, had no well-developed homing instinct, were no migratory, and did not need to rendezvous in special places in order to find mates. Polygyny had not developed in any of the pinniped stocks. The animals were perhaps smaller than most recent pinnipeds had less fat, lived in more temperate waters, and perhaps made crude dens in beach grasses and among boulders. The advancing pinnipeds met, from sea birds and cetaceans, little competition for food. They met no competition for breeding room, nor do they often today. Pinnipeds early lost the habit of feeding on beach organisms such as crabs, mussels, periwinkles and blennies.

As each generic stock became isolated, it was transformed in response to the local physical and biotic environment and adapted to its peculiar niche. Among species frequenting isolated bays, gulfs, islands, inland seas, or lakes, there has been and still is, rapid evolution. In the last few centuries, the effect of commercial sealing upon island and continental pinniped populations has been catastrophic. Millions of seals have been slaughtered and entire breeding populations wiped out. It is doubtful that some species will ever recover genetically from under-going such a severe population "bottleneck". In one recently exploited species, absolutely do polymorphic variation was found in 21 blood proteins from 159 seals representing five rookeries! In this regards, some land carnivore populations have suffered a similar fate at the hands of man-for example, the wolf. The increasing range of human populations is changing the distribution of both land carnivores and pinnipeds.

Inspite of the fact the pinnipeds are marine, all species have retained an attachment to land. In this respect, they differ from the Cetacea, descendants of a primitive ungulate ancestor, who have become completely aquatic. All pinnipeds give birth to their youngs on land or ice, and some species spend the majority of their time resting out of the water. Pinnipeps have exploited the marine niche for food. A few carnivores, such as the sea otter, Enhydra lutris, and the polar bear, Thalarctos maritimus, also obtain food from the sea, but they have not undergone the same degree of morphological transformation to aquatic living as the pinnipeds. The anatomical and behavioural adaptations of the pinnipeds to life in the sea clearly distinguishes them from the land carnivores. Since these adaptations influence their behaviour on land, it is worth reviewing some of the major ones.

MORPHOLOGICAL COMPARISONS

(a) Size

There are several morphological differences between pinnipeds and land carnivores simply reflecting the different environments in which they live. The average body size of pinnipeds is greater than that of carnivores. The smallest pinniped, the ringed seal, pusa hispida, weighs 90 kg. The largest pinniped, the southern elephant seal, Mirounga leonina, weighs 771 kg. about 3629 kg and is 650 cm long. Several seal species are larger than the largest land carnivore, the grizzly bear, Ursus arctos, which weighs. Many land carnivores, such as the foxes of Africa weigh less than 4 kgs. The Fennec fox, Fennecus zerda, weighs less than 1 kg! The greater body size of pinnipeds evolved mainly in response to the marine environment. Large body size is better for heat retention in the cold sea. Moreover, since the water medium gives more support than air, large size could more easily evolve in the sea than on land.

(b) Shape

Aquatic adaptation favouring swimming and diving have been observed in pinnapeds. Body shape is streamlined to reduce drag. The external ears are reduced or absent, external genitalia and mammary tests are drawn into the body, limbs are enclosed within the body and extremities are flattened—the tail is short and the head is flattened, and the eyes are situated well forward. The neck, in particular, is thick and muscular and considerably more flexible than in most large carnivores. The skin also shows adaptation to a water environment, and the hair is flattened. Pinnipeds never groom the pelage with the mouth or tongue. The terrestrial carnivore body is a picture of contrast. The canid body, especially the large one, is designed for long-distance running.

(c) Subcutaneious Fat

Subcutaneous fat forms a substantial part of all pinniped bodies. Skin and blubber make up more than 25 per cent of the weight of the Weddell seal, Leptonychotes weddelli and almost 50 per cent of the body weight of the southern elephant seal. Blubber the anoetified adipore layaprovides reserve energy during fats and lactation, thermal insulation, buoyancy, and the padding necessary for a streamlined profile. The proportion of body fat to total weight in land carnivores is apparently much smaller, even in dormant or hibernating bears, skunks, and raccoons.

(d) Brain

Although the pinniped brain has not been studied in great detail, it is evidently large, like that of terrestrial carnivores, but more spherical and more highly covoluted. The sea lion brain resembles that of bears, while the seal brain looks more like that of cats and dogs. Compared to a dog's br ain, the cerebellum is large, probably because of the increased coordination demanded in swimming. The auditory nerve is large and the olfactory lobes are reduced. Pinniped eyes (the walrus excepted) are larger than those of land carnivores, and they function well at low levels of illumination. Seal vision is excellent both in water and on land.

(e) Teeth

Pinnipeds have fewer and more uniform teeth than most land carnivores (except for the walrus, a special case once again). All species have the pronounced upper and lower canines but lack the carnassical cusps characteristic of land carnivores. The postcanines in most species are rudimentary pegs functioning to hold prey that is swallowed whole. Variation in pinniped teeth reflects the type of flesh the animal feed on, and, as in many mammals, it is a useful taxonomic indicator. Generally, the teeth, mouth, jaw, and associated structures of pinnipeds are designed for grasping, tearing, and swallowing prey whole or in chunks, rather than for chewing, shearing, or crushing large bones. In contrast with land carnivores, the deciduous teeth or milk teeth of pinnipeds disappear before or soon after birth.

Several physiological adjustments to aquatic life have been made on pinnipeds that set them apart from land carnivores. These adaptations in respiration, circulation, renal psysiology, and the like are important and interesting, but it would be too much of digression to cover these topics here.

PREY, FOOD HABITS, AND FEEDING BEHAVIOUR

Pinnipeds deviated from land carnivores in form and function because of adaptations that took place as a result of seeking food in the seal. It is therefore most important to compare these two animal groups from the point of view of hunting and feeding habits, and in relation to their respective prey.

(a) Prey

All carnivores including pinnipeds and land carnivores are meat eater, but only the marine mammals eat flesh exclusively. Seals and sea lions feed mainly on crustacea, molluscs, fish, penguines, and other sea birds. Although diet varies with the species, the majority of pinnipeds feed on a variety of prey. The Alaska fur seal, Callorhinus ursinus, is the best studied pinniped from the point of view of food habits. Its catholic diet is representative of the wide range of fishes and squid eaten by the Otariids (the earred seals or sea lions and fur seals, as opposed to the Phocids, the earless, true seals). Alaska fur seals feed on more than 30 kinds of marine organisms: at least 27 species of fishes, 1 species of octopus, and 5 species of squid. This most common prey of fur seals are northern anchovy, Engraulis mordax, squid, and Pacific herring. (Clupea harengus). Various rockfish, Pacific hake, Merluccious products, Pacific saury, Cololabis saira, salmon, Oncorhynchus appeared Americans had, Alasa sapidissimia, are eaten in lesser quantities. Hermit crabs, amphipods, and various diving sea birds (*e.g.*, Rhinocerous Auklet, Cerohinca monocerata, Redthroated Loon, Gavia stellata, and Best Petrel, Oceanodroma leucorhoa beali) are eaten occasionally, but they seem to play only a minor role in the seals' diet.

Fur seals, like most other pinnipeds, are opportunistic feeders. What is eaten depends on the seasonal distribution and the abundance of prey. Since fur seals commonly feed on schooling fishes, stomach contents of several animals collected in the same area often contain only one type of fish. The Steller sea lion, Eumetopias jubata, and the California sea lion, Zalophus californianus, eat similar fishes and squid, especially where the feeding range of the three species overlaps. Steller males in the Bering Sea and off Alaskan islands supplement their diet with Alaska fur seal pups. Most true seals, such as the harbor seal, Phoca vitulina, and the grey seal, Halichoerus grypus, are mainly fish eaters, but they occasionally feed on an assortment of crustac, octopus; eels, and molluscs. The northern elephant seal, Mirouna angustirostris, is a deep diving seal that feeds primarily on squid, skates, rays, small sharks, and ratifish, Hydrolages collier. The small riuged seal, Pusa Hispida, which inhabits the circumpolar Arctic coasts, feeds on up to 72 different species of small pelagic amphipods, euphausians, and other crustacea, as well as on small fishes.

Only two of the 32 different pinniped species are rather specialized feeders that exploit one type of prey species almost exclusively. The Carbeater, Lobodon carcinophagus, feeds on krill, small shrimp-like animals that it catches in quantity and strains from the water through its multicusped cheek teeth, much as a mysticete whale sieves krill through its baleen. The walrus, Odobenus rosmarus feeds primarily on three genera of bivalve molluscs, Mya, Saxicava, and Cardium, for which it forages in shallow coastal waters less than 40 fathoms deep. Bivalves on the sea bottom are examined and sorted by the lips and whiskers; feet and fleshy parts are torn off and swallowed whole or sucked out. But when molluscs are scarce, the walrus may eat young ringed seals, bearded seals, Erignathus barbatus, even young walruses. It is the only pinniped that aparently feeds on Cetacea. Occasionally a narwhal, Monodon monoceros, or a beluga, Delphinapterus leucas is eaten. However, it is not known whether walruses acturally kill them or feed on corpses.

According to Mech "probably every kind of backboned animal that lives in the range of the wolf has been eaten by the wolf." Wolves, Canis lupus, eat mice, mink, muskrats, squirrels, rabbits, various birds, fishes, lizards, snakes, grasshoppers, earthworms, and berries. But predation on small animals plays only a minor role in the wolf's diet. The main prey of the wolf, like those of the African wild dog, Lycaon pictus, are large animals. The wolf's primary prey in the United States and Canada are: white-tailed deer, mule deer, moose, caribou, elk, Dall sheep, bighorn sheep and beaver. Wild dog kill mostly Thomson's gazelles, juvenile wildebeests and Grant's gazelles and, occasionally, warthogs and zebras. African foxes feed mostly on rodents and small reptiles, bird's eggs, insects and vegetable matter; jackals eat similar foods in addition to small mammals and carrion. Additional information on other canids can be found in Fox.

Of all the land carnivores, bears have the most diverse tastes. The grizzly, before man virtually annihilated it, was an omnivorous opportunist part excellence. The grizzly ate almost anything and everything that was available. This long list included: meat, fresh or putrid from whales, water birds, fishes, elk, deer. antelope, gophers, lizards, frogs, and domestic livestock; a wider variety of plant materials than herbivorous animals ate, some of which were various berries, clovers, nuts, wheat corn, potatoes, tree bark, and various bulbs and assorted foods like honey, ants and their larvae, yellow-jacket nests, and mushrooms. Polar bears, Thalarctos, maritimus have a much more restricted diet. Although they consume some vegetation and carrion during the summer, they feed primarily on the ringed seal during most of the year.

Size of Prey

Among large carnivores there is a remarkable thing about their feeding habits is that the prey is often larger than the predator. In this respect, the large canids differ from all of the pinnipeds. Wild dogs, whose average weight does not exceed 18 kg, bring down impala and reedbucks that are double or triple their size and zebras that are even larger. Wolves weighing 36 to 45 kg kill moose and bison that may weigh over 500 kg. These canids overcome large prey by hunting in packs that range from a few animals up to a score or more of them. One dog usually selects a quarry from a retreating herd of gazelles, and the other dogs follow it. The quarry is coursed for 1 to 3 km until it becomes exhausted. The lead dog catches up to it and grabs it or bowls it over. Once overtaken, the prey is fallen on by all the following dogs, and it is dismembered and eaten with great dispatch. Wolves bring down deer or moose in a similar way. The first wolf to catch up to the fleeing prey attempts to grab hold of the rump, flanks, neck, or nose. Once the prey is down, other attack from every side. Although the technique may vary with prey species, the strategy of the individual quarry, and the composition of the hunting pack, it is clear that these pack hunters get help from each other in getting their food. Lone wolves or even small packs are at a disadvantage in bringing down large prey. In black-backed jackals, Canis mesomelas two hunters are more than four times as successful as one in bringing down gazzelle fawns as observed by Wyman.

In pinnipeds the size of prey eaten varies greatly like land carnivores, but, except for the dobious possibility that walruses kill small whales, pinnipeds always kill and eat animals that are much smaller than themselves. Prey size determines how alaska fur seals, Steller sea lions, and harbor seals eat their food. Small fishes less than 30 cm long, such as anchovy, herring, saury, or squid, are consumed whole under water. Except for very small food items like lanternfish, the prey is swallowed head first. Larger prey measuring more than 30 cm long, such as hake, rockfish, or salmon, is brought to the surface, grasped by the head, shaken violently, and reduced to chunks that are swallowed piecemeal. As fur seals get older and larger, large fishes up to 92 cm make up an increasingly high proportion of their diet. Thus the size of the prey increases with the growing age of fish seas.

In most of pinnipeds the hunting is done in groups, but, unlike the case with canids, grouping does not enable them to exploit larger animals. Rather, grouping seems to make it easier for them to exploit large aggregations of small prey. Fiscus and Baines saw Steller sea lions leaving their hauling grounds in compact groups of several hundred to several thousand animals. They swam out to feeding areas, where they dispersed into smaller groups of less than 50 animals containing both sexes and mixed sizes of animals. Massings of this type fed on large schools of fish or squid. Casual observations suggest that groups of sea lions feed more efficiently on schools of fish because cooperation among the hunters enables them to herd and control the movements of the school. When large fish schools are absent, sea lions feed singly or in small groups of two to five animals.

In some cases the hunting is performed in a very special way before Steller sea lion females leave the rookery and their pups to go to sea and feed, they may engage in activities that resemble, and that may have an analogous function to, the prehunt greeting ceremony of wild dogs. Several females may gather at water's edge, where they mill about restlessly in close contact with each other. They may vocalize repeatedly and engage in what appears to be low-intensity aggressive behaviour before swimming off as group to feed. In both species, these activities may reinforce group cohesion and unity as well as synchronize the time of departure.

Similar observations have also been made in canids and pinnipeds. They may travel several kilometers to get a meal. Van Lawick-Goodall and van Lawick-Goodall report that wild dogs did not start their first chase until they were 8 km from where they started. The chase itself covered a distance of $5\frac{1}{2}$ km. Wolves may pursue moose or caribou for up to 5 to 8 km but usually give up sooner. They may range up to 32 km from their den. Steller sea lions and California sea lions travel much further from their rookeries or hauling grounds to feed. The former have been seen feeding as far as 112 to 136 kms from land and the latter as for as 62 kms from land. Usually, only small groups of less than 30 individuals are seen this far from land. Large groups of sea lions (100 or more) are seldom seen feeding more than 16 to 24 km from a hauling ground or rookery. It would be interesting to compare land carnivores and pinnipeds on the success of their respective hunting and fishing expeditions, but the difficulties inherent in observing seals feeding in the water makes this impossible at present. However, some indirect evidence is pertinent.

Although both pinnipeds and the larger land carnivores are nomadic, the former travel greater distances in their annual feeding migrations. The record holder is the Alaska fur seal, which migrates from its breeding grounds in the Bering Sea to its winter feeding grounds along the shores of the eastern and western Pacific as far south as san Diego, California, and Hokkaido, Japan, a distance of some 5000 km. Wolves may follow caribou on their annual migrations.

Times of Hunting

Wild dogs and dholes, Ceron alpinus; prefer to hunt in early morning or early afternoon or evening, although hunts occur on moonlight nights. Wolves and bears may hunt by dry or night. Most of the smaller foxes are nacturnal feeders. Fur seals and sea lions are primarily night and early morning feeders. Stomach contents of animals collected at various hours after sunrise show decreasing amounts of food. However, in some areas where large schools of fish are present, daytime feeding also occurs. Typically, the midday hours are spent resting, and feeding activities gradually resume in late afternoon. The harbor seal is a daytime feeder.

Amount of Food

The amount of food varies for animal to animal. Canids eat great amounts of food in a single feeding bout. A wolf may gorge 9 kg of meat in a singly meal; this may represent 20 per cent of its body weight. However, estimates of a little more than half that amount per day per wolf may be more representative. Estes and Goddard estimate that wild dogs average 2.72 kg of meat per day, or approximately 6.7 per cent of their body weight. But these canids may have to go without eating for several days at a time. Five to 7 day fasts have been recorded in wolves, and one wolf went for 17 days without food. It is common knowledge that domestic dogs can go for several days without eating, particularly when sexual activity is probable. Grizzlies and black bears in the more northerly parts of North America may go dormant for varying periods of time during the winter when food is scarce. In general, all of the land carnivores will feed regularly if possible. Even bears will actively feed throughout the year when food is available.

Fasting

Among aquatic a carnivool mammals the pinnipeds feed daily except at certain times of the year, usually the breeding season, when they undergo long fasts. Thus, feasts and famine are even more extreme in their annual cycle. When they are feeding, fur seals and sea lions ingest less food relative to body weight than the canids mentioned above. Spaulding estimates that fur seal, seal lion, and harbor seal food requirements range from 2 to 11 per cent of body weight per day, with 6 per cent being the average. During the breeding season, fur seal and sea lion males fast for approximately 6 weeks; harbor seals do not appear to fast during this time. The record for long fasts in pinnipeds goes to male northern elephant seals. They may go without food for 3 months. Throughout this time they are actively fighting and attempting to copulate, and they may lose over 450 kgs. Females of this species fast for an average of 34 days during the period in which they give birth and nurse their pups daily for 28 days. A female's total body weight may be reduced by almost 50 per cent during this fast. During the nonbreeding season, males and females fast again for approximately 30 days while undergoing the annual molt.

Specific Feeding

Most canids and ursids scavenge for food at sometime. Perhaps bears exploit carrion more than any other land carivore. Carrion was a major part of the California grizzly's diet more than 100 years ago, before man killed the grizzly and its prey. It is said that these bears kept the California coast clean of pinniped and whale carcasses that repeatedly washed ashore. There are numerous reports of 12 to 15 bears feeding at the same time on a single whale carcass. Many of the larger land carnivores typically desert the remains of a kill and cache it away and return to eat from it later. Polar bears are an exception to this rule. They no not cache ringed seal carcasses that they kill, despite the fact that they may only feed on the blubber and leave the meat behind. In marked contrast to these feeding habits of land carnivores, pinnipeds do not cache food, nor do they scavenge, although the walrus may be an exception to this rule. Pinnipeds consume most of their prey entire and discard only the heads of large fish, particularly rockfish.

Predation

One drawback that pinnipeds experience in their marine habitat to which the larger land carnivores are relatively immune is predation. In carving their niche in the sea, pinnipeds became exposed to large predators who were also adapted to aquatic life and more formidable than themselves. At sea, the main predators on pinnipeds are large sharks, particularly the great white shark, Carcharodon carcharias, and the killer whale, Orcinus orca. All pinniped species living in the Pacific are preyed on by these animals. The degree of predation is unknown. When pinnipeds return to land or ice to give birth and nurse their young, they are vulnerable to predation by large land carnivores such as polar bears and grizzly bears, perhaps wolves, and, of course, man. Small carnivores, such as foxes, and skuas gulls, and hawks, may prey on their young. Being adapted primarily for locomotion in water, they are no match defending themselves against large land predators. Thus, it would appear that predator pressure was important in causing pinnipeds to breed in remote sanctuaries free from these predators; that is, offshore islands or rocks, sandbars and ice floes, and offshore breeding areas are limited.

SOCIAL BEHAVIOUR AND REPRODUCTION

It is likely that pinnipeds were forced to share limited breeding areas, and this led to the high degree of sociality and the large social gatherings that we find in pinnipeds today. In terms of the sheer number of animals that congregate together, the pinnipeds are much more social than any land carnivore. As many as 2 million Alaska fur seals can be found every summer on two tiny Pribilof Islands in the Bering Sea. Three thousand northern elephant seals may be found packed together tightly on one beach during the peak of the breeding season. Two thousand California sea lions may be found sleeping together in close contact during the nonbreeding season. The largest wolf or wild dog packs are miniscule in comparison.

Pinnipeds are among the most polygynous mammals. Apparently, this mode of life developed very early in their history, and it was closely tied to their amphibious habits. When females began clustering in time and space, the conditions became ideal for promoting male-male competition, polygyny, and sexual dimorphism. Males must have competed to mate with as many females as possible. Those who kept other males away from areas containing females sired more pups. In their short breeding seasons on traditional rookeries, fur seals and sea lions developed a social structure characterized by territoriality among males. Males without territories do not copulate, or they copulate only rarely; those who secure the most well-placed territories – territories containing the most estrous females—do most of the breeding. Another system that developed was for males to dominate other males and thus gain access to more estrous females wherever, they were situated. This was the strategy adopted by northern elephant seal males. They fight for social status in a dominance hierarchy. The highest ranking male or males-depending on the number of females present—locate themselves nearest the females and keep all other away. One or a few of the highest-ranking males do most of the breeding in a harem; most males in a colony do not breed at all. One male may dominate a harem for several years and inseminate more than 200 females. Grey seal society is similar to that of elephant seals.

Thus, in most of the well-studied pinnipeds, males are polygynous and are either territorial or exhibit a dominance heirarchy during the breeding season. This is a reflection of the manner in which the largest and strongest males go about monopolizing a large number of females. Since size and physical aggression convey such a great reproductive advantage to males, these traits have become increasingly elaborated in time. Sexual dimorphism is greater in some seals and sea lions than in other mammals. For example, the Alaska fur seal male is five time larger than the female.

Among terrestrial carnivores the Canids are "social" in quite a different way from pinnipeds. Social organization is characterized by seasonal or permanent pair bonds in foxes and jackals, and by cooperative hunting and group life in the wolf, dhole, and wild dog. Black bears are essentially solitary, but female-cub associations are long lasting, and strong and female cubs inherit their mother's feeding territories. Wolf packs are organized into separate male and female social hierarchies. Although the female hierarchy is more ambiguous, it is clearly more pronounced than the size-related dominance observed in female elephant seals. Hunting dog packs have similar hierarchies, particularly during the breeding season, but it is female hierarchy that is most evident. In wolves and wild dogs, the dominant breeding female may prevent subordinate females from breeding or may drive them out of the pack and kill their pups. Dominant females may also enlist the help of males, and they may prevent other adults from feeding the subordinate female or her young.

Territotrial Behaviour

Although many canids and ursids are territorial in the sense that they defend a feeding or home range against other individuals or groups of individuals, this behaviour is quite different from the defense of rigid individual territories by fur seal males and Steller sea lion males against conspecific males during the breeding season.

Sexual Behaviour

When it comes to reproductive behaviour and parental care, pinnipeds and terrestrial carnivores differ in many ways. Most canids and ursids, like pinnipeds, have one estrus per year. However, male-female sexual interactions in the land carnivores can be prolonged and complex. Canids exhibit extensive precopulatory courtship, frequent olfactory examination of the partners, and frequent urine and faecal marking. In some domestic canids, females actively solicit males. Canid copulation is characterized by a genital lock or tie. After intromission is achieved, the pair remains joined together by the genitalia and face in opposite directions for several minutes or as long as an hour. Females may be in estrus for several days and may copulate several times with different males. Usually, only the alpha female in a wolf pack breeds.

In contrast, there is little or no courtship in pinniped sexual behaviour. Male northern elephant seals pounce on females and attempt to force copulation at any time during the breeding season, regardless of the female's reproductive condition. Sea lions and fur seals may investigate the femal e's genitalia before mounting, but there is little parallel to the direct solicitation of males by females seen in the domestic dog. Male seals do not mark with urine or feces, and they do not exhibit a copulatory lock. Copulation duration is equally long in both animal groups, lasting 3 to 7 min in elephant seals and 15 to 20 min in Steller sea lions. Some phocid females are in estrus for several days; most otariid females (in which the males are territorial) copulate only once during the estros period. Copulation is more canem except in elephant seals, where the male mounts from the rear and to one side, perhaps a necessity because of the large bulk of the animals. Elephant seal males, like male wolves, prevent or interrupt each other's copulations; female seals, unlike wolf bitches, do not prevent other females from mating, but they threaten, attack, and on occasion kill pups of neighbouring females.

Gestation and Parental Care

Canids have a short gestation period of about 2 months and a litter size that ranges from 2 to 15, with 5 being the mode. The gestation period of bears is 6 to 7½, months, and the litter size is usually 2 to 4. Wild canid pups are born in dens or burrows. Weaning from mother's milk is at about 40 days in the domestic dog. But during the transition to solid food in most canids, the altricial young are provisioned by regurgitation feeding of partially masticated food. This is done by the parent male as well as by the mother in the smaller pair-bonding canids. In wolves and wild dogs, all adults in the pack, male and female, regurgitate to young pups as well as to the nursing mother not involved in the hunt. It is not clear how long this type of provisioning of the young goes on, but apparently it lasts until the young are able to obtain food on their own.

Females of all pinniped species give birth to a single pup once a year. They do not eat the placenta as the carnivores do. Within a few days or weeks after parturition, copulation takes place. Gestation, including delayed implantation, last 1 year. Thus, females give birth at about the same time each year and usually in the same place (except for seals that breed on ice). Most species give birth in the open, whether it is on sand, rock or ice, the ringed seal is an exception and gives birth in subnivean lairs in the pressure ridges of sea ice. Newborns of all species are physically well developed at birth. Pups are nursed for a minimum of 3 weeks to more than a year. The northern elephant seal female is a mammalian phenomenon in that she doesn't feed during the entire 4 week nursing period! Weaning is abrupt in species like the elephant seal and gradual in others like the Pacific sea lions. Apparently, mothers feed their pups exclusively by nursing; they have not been observed carrying food to the young or regurgitating to them. Females to several species that breed colonially will not nurse pups other than their own. Northern elephant seal females bite and sometimes kill orphaned pups that attempt to suckle them. Adoption of orphaned pups by females who have lost their own pups may occur in elephant seals, grey seals, and, rarely, in Steller sea lions. Unlike the canids, male pinnipeds that breed in colonies do not help in rearing or feeding the young. Quite the contrary, males are inadvertently responsible for a sizeable portion of pup mortality in elephant seals and grey seals.

Evidently, there are fundamental differences in the social and reproductive behaviour of these two related groups of animals. Two basic differences having to do with the origin of sociality and the rate of reproductive behaviour stand out.

Social behaviour seems to have quite a different origin in these two animal groups; that is, they must have been exposed to quite different selection pressures. In the carnivores, there developed a premium on grouping together, cooperating in the hunt, and feeding other pack members, because these behaviours increased the efficiency of predation. Being social enhanced survival through the process of kin selection. On the other hand, pinniped sociality seems to have developed first because of a paucity of breeding sites free of predators; later, this trend became elaborated through the operation of sexual selection. It is unique to pinnipeds that feeding and breeding have always been spatially separated and have occurred in different habitats. Unlike the typical terrestrial carnivore with an extensive home range in which mating also takes place, pinnipeds opted to travel great distances to breed on traditional sancturies. Since such sancturies were limited, females were forced to clump together; this situation predisposed males to maximize their reproductive success by attempting to inseminate as many females as possible. Successful males were belligerent and big. Unbridled male-male competition led quickly to extreme polygyny, sexual dimorphism, lack of parental behaviour on the part of the male, sex differences in behaviour, and continued gathering of a large number of animals for breeding. Once this system was started, the inertia of sexual selection kept it moving in the same direction. For example, females that move from the main aggregation of females are apt to be fertilized by a male who is unsuccessful in competing with other males—that is, a "marginal" male. If he is subordinate because of genetic factors, his progeny will be less successful in the next generation. Thus, any tendency of females to become less gregarious would be selected against and would be self-limiting. Gregariousness in females should be positively reinforced as long as competition among males—whether for territories or for social status—results in the exclusion of some males.

The second fundamental difference between pinnipeds and terrestrial carnivores has to do with reproductive turnover. In general, natural selection have favoured two contrasting responses to the problems of population replacement and the colonization of new habitats. According to MacArthur and Wilson, in environments where there is little crowding and food is abundant, r-selection will favour animals that utilize the most food and rear the larger families—this is, those with a high population growth rate. On the other hand, when there is crowding and intense competition for food, K-selection will favour animals that can replace themselves with the lowest possible intake of food. The first is a strategy of productivity, and the second is one of efficiency of reproduction.

In general, pinnipeds appear to have been K-selected and terrestrial carnivores r-selected-Pianka's admonition that all organisms represent a compromise on this continuum is certainly applicable here. One of the most interesting correlates of γ-and K-selection is the pair of complexes that Martin calls, respectively, the altricial and precocial complex. Pinnipeds conform to many of the attributes of the precocial complex. Only one pup is born per year and the young exhibit advanced physical development at birth. Alaska fur seals can locomote and vocalize within a moment of birth, and they are as advanced at birth as domestic dogs about 3 weeks old. Northern elephant seal pups open their eyes and may exhibit the characteristic sand-flipping movements of the foreflippers before completely expelled from the birth canal. Harbour seal pups are able to swim and dive at birth. Lactation is of long duration, lasting more than year in some sea lions and the walrus. Even in seals having a relatively short nursing period, energy investment in the offspring may be enormous. Elephant seal milk is approximately 55 per cent fat, and although pups are weaned at 4 weeks of age, they are left with a tremendous fat store. These pups weigh more at weaning than they do a year later! Other attributes of the precocial complex are late sexual maturity and a long life. Most pinniped females ovulate for the first time at about 2 to 5 years of age; males reach sexual maturity at about 3 to 6 years of age harbour seals, Alaska fur seals, anonymous, northern elephant seals. In the colonial species, males do not complete successfully for females until much later in life, until approximately 10 years of age in the northern elephant seal. Pinnipeds may live a long time. Harbour seals live for 20 to 29 years. Alaska fur seals for up to 26 years, and the life span of the northern elephant seal is typically about 14 years. Like that in primates and other long-lived mammals, the trend in pinniped reproduction has been toward quality rather than quantity.

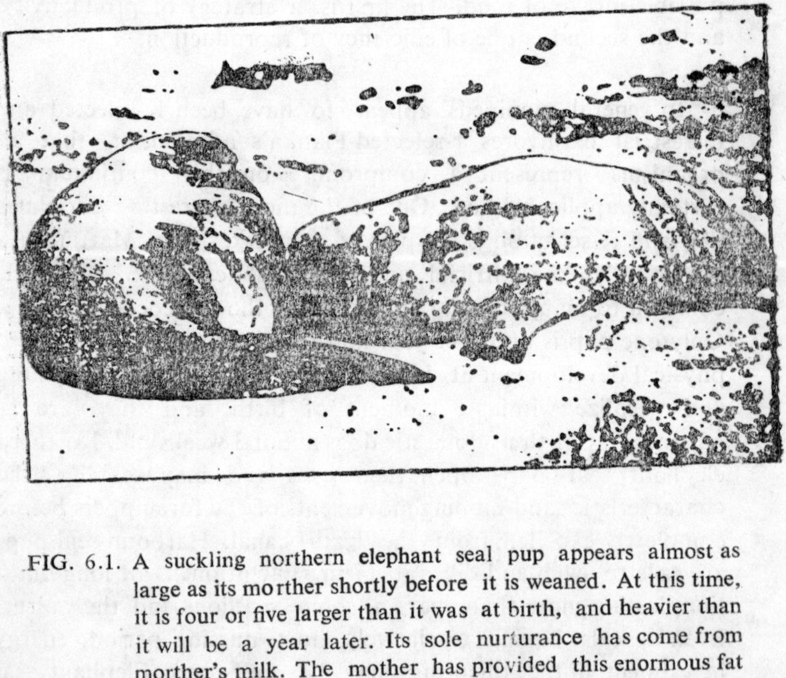

FIG. 6.1. A suckling northern elephant seal pup appears almost as large as its morther shortly before it is weaned. At this time, it is four or five larger than it was at birth, and heavier than it will be a year later. Its sole nurturance has come from morther's milk. The mother has provided this enormous fat store to her offspring during a four-week period in which she did not feed.

Terrestrial carnivores, especially the smaller canids, exhibit the altricial complex. Most canids give birth in a shelter to a large litter of pups that are poorly developed and incapable of coordinated behaviour except suckling. Gestation is short and lactation brief compared to most pinnipeds. Sexual maturity occurs at an early age, at the end of the second year in wild wolves, and at the end of the first year is some captive wolves and in the Arctic fox. Life spans are short compared to those of pinnipeds. Although wolves have lived 10 to 16 years in captivity, 10 is considered very old. Seven is considered old age in large dogs. The life span of foxes is considerably less, 6 years of age in the gray fox, Urocyon littoralis. Apparently, canids and ursids have been selected for quantity in reproduction. The paradoxically low fecundity rate in wolf and wild dog packs, where only one female may give birth, may be related to low availability of food. It is interesting to note that if only one female gives birth in a pack, it is usually the most dominant one. One might expect the fecundity rate to go up when food was abundant. Whether this happens or not, the point is that the potential for a high reproductive rate is there.

The pinnipeds must deal with problems that arise from making a living to two quite different habitats. The high metabolic rate and thick layer of blubber that are so adaptive in cold water create difficulties in thermoregulation on land, particularly when the ambient temperature is high, or after sustained activity has taken place. Numerous behavioural adjustments are important in preventing overheading. In addition, one negative consequence of the large social gatherings on land is that the animals are more vulnerable to disease and parasite transmission. The colonial breeding sea lions contain a wide variety of internal and external parasites. For example, seven species of trematodes, eight species of nematodes, one cestode, and various species of Acanthocephala. Acarina, and Anoplura have been identified in the California sea lion. Lungworm infestation is particularly heavy in this species and may be an important cause of juvenile mortality. Land carnivores do not seem to harbour the variety or the same high levels of parasite infestation as do these marine mammals.

7

Human Behaviour

Of all the organs of the body, by far the most complex is the brain. The evolution of the intricacies of the human brain has enabled Homo sapiens to become the dominant species on the planet Earth. The human is the animal which is best equipped to overcome challenges imposed by the environment. Largely as a result of his superior mental abilities, the human is less restricted than other species by environmental problems. In this sense, his greater independence of environment, the human is the most highly developed of all creatures. Not only does he cope with existing environmental problems, but he is also capable of changing the environment to suit his needs. In so doing, he is also influences the course of his own evolution as well as that of other species. The ultimate benefits to be reaped from tempering with natural selection poses many profound questions bearing on the very future of life on earth. How the human uses his superior brain to manipulate the environment is a direct reflection of human intelligence. Intelligence is reflected in the many facets of human behaviour which has the potential to influence the very existence of all life forms.

Any information which can be obtained to provide an insight into the nature of human intelligence and behaviour has a decided value. Since the human nervous system is so complex, it should come as no surprise that analysis of the hereditary and environmental components which influence human intelligence and behaviour is the most difficult of all genetic investigations. However, many facts related to the problem are known. Sufficient data may in time accumulate to permit a breakthrough into a fuller appreciation of the human mind and the reasons why the human acts as he does.

Let us begin by noting several obvious facts about one aspect of human behavirour, speaking a language. The ability to speak entails a decided advantage for the species. It enables members of one generation to pass the assembled knowledge, the culture of the society, to the next generation with no less of time and no need on the part of the later generation to learn everything anew. Without this ability, each new generation of a species loses the advantage to profit from the knowledge accumulated by the efforts of previous ones. The ability to communicate through speech and the written words associated with oral language makes it possible for each generation to build on the culture of its ancestors, to utilize it, and improve it, thus becoming more and more independent of environmental problems.

These points about language indicate something about behaviour in general. Just like any other feature of an individual, behaviour is a phenotypic characteristic. We speak of species-specific behaviour when we refer to the assemblage of behaviour patterns which is considered typical of a given animal. We expect members of a given species to behave or conduct themselves in a certain way. A particular species of bird goes through its expected courting behaviour, builds and nest incubates its eggs, and feeds its young. The young, in turn, respond to the mother in a typical fashion. It the bird is a duck, the young will follow her shortly after hatching and will soon learn to swim.

The behaviour of human parents in rearing their young has specific features which are expressed during the long period of development required by the very helpless human infant. While incapable shortly after birth of performing feats such as walking, swimming, or feeding himself, the potential of the human infant exceeds by far that of any other animal. The human's genetic endowment has provided him with a central nervous system of the utmost complexity. The countless nerve connections and relays in the brain itself provide him with the machinery to accomplish something no other animal can do. This is the ability to deal with abstractions. To receive symbols, arrange, and interpret them in a meaningful coherent fashion. There is evidence from studies with chimpanzees that these animals possess a certain ability to deal intelligently with symobls, but it does not approach the facility of the human in this respect. Moreover, the human is unique in the ability to communicates symbols freely through a spoken language. Again, chimpanzees have been taught after labour to utter a few simple words. The inability of the ape to speak freely may not be entirely a reflection of limited intelligence. It is to a large degree a consequence of the fact that the ape is not equipped anatomically to handle spoken words. Speech demands a voice box, palate, nose, and mouth cavities of a certain construction. No other animal possess the type of anatomy required for the free formation of words.

Also essential for proper speech is a normal hearing apparatus. This is the receiving part of the system which delivers the sound stimuli to the brain. It is here that these stimuli are arranged and interpreted into a meaningful whole. Before any human can speak properly, he must be provided with these sound stimuli. These enable him in turn to form his own sound by association of symbols in the complexes of his brain. Ability to speak a language, therefore, is a complex characteristic. It depends in large part on anatomy, the hearing apparatus to receive the sound stimuli, the brain cells to interpret them, and the vocal apparatus to form words. The anatomical features necessary for language communication have a large hereditary component. Certainly, every fertile human must carry in the sperm or egg information required for the embryo to develop these anatomical requirements. A defective gene may cause upsets in the differentation of one of these structural parts. Without sound reception, a human will never be able to speak clearly and freely, even though the brain and vocal apparatus have developed perfectly.

But proper hereditary information is by no means the sole requirement. An environmental trauma (viral attack during fetal development, birth accident) may deprive an infant of sound reception. Again, without the appreciation of sound, even though the proper genetic endowment is present, such a person will lack the facility with spoken words possessed by a more fortunate individual.

The ability to deal with symbols is a reflection of the intangible known as intelligence. The increase in intelligence which the human possesses over other animals is seen in the greater ability of memory retention, to think, and to arrange information stored in the brain in such a manner that foresight is possible. This foresight enables the human to act in ways which permit him to avoid potentially harmful environmental influences. With foresight, the human can plan and transmit his knowledge and culture. This thinking, learning and foresight depend, of course, on a brain with billions of neurons (nerve cells). Increase in intelligence thus depends to a large degree on increase in the number of brain cells and thus increase in the size of the brain. It is well known that differences in intelligence among different animal species is related to differences in brain size. However, within a species, there is no evidence to indicate that a larger brain reflects greater intelligence. Nevertheless, the species-specific behaviour of the human in regard to the ability to think, plan and speak depends on an important genetic component which provides the information to construct a brain of a certain size as well as the other anatomical features essential for the expression of the brain's full potential. However, the environment in which these genetic factors operate is a paramount importance may carry genetic factors for normal speech or above-average intelligence, the environment can deprive him of the full benefits which they provide. We see that behaviour, like any other phenotypic characteristic, it thus the result of the interaction of many genes and the environment. It should also be evident that species-specific behaviour has been subjected to the operation of the forces of evolution. Natural selection has provided each species with a particular behaviour which enables it to cope best with its environmental challenges. Behaviour, therefore, has an advantage just as any other characteristic of a species of individual. The human's behaviour, as noted here, has enabled him to become less restricted than any other species, to be able to make use of the greatest number of environmental niches, and to manipulate the environment for his own end.

While much of human behaviour depends on heredity, it must not be forgotten that a great deal of it depends on learning. Human behaviour is distinguished by its plasticity, its ability to be changed and molded. A normal child is born with genetic factors which make it possible for him to receive environmental stimuli. Among the most important of these are the ones provided by other humans. An infant is associated with a family, and as he continues to develop, he is exposed to more and more members of his particular cultural group or society. Certain stimuli will determine what language he will speak. He carries the genetic information which provides the ability to speak, but whether he will speak and what language he will become expert in depends on other humans about him. The amazing inherent talent which a child posseses to learn a language is seen by the youngste's ability to become expert in several tongues if he receives the appropriate sounds during his formative years. No one must tell him what is English, Chinese, or French. He sorts them out automatically without being told a thing about rules of grammar. But alas, this genetic ability is lost about the age of puberty, so that no adult will ever have the facility of the child who can so effortlessly master two or more tongues. No will most adults ever speak a language perfectly if they have learned it after they have reached sexual maturity. There is nothing that dictates what language a child must speak. The great plasticity of this aspect of human behaviour is quite apparent, as is the interaction of its hereditary and environmental components.

Few human behavioural traits are based exclusively on genetic factors. The plasticity provide by the environmental interaction with the genetic component is seen in a child's general conditioning as he becomes integrated into his particular society. He learns the customs of his society and tends to behave according to the expectations of his specific cultural group. Reared in a completely different society with different values and expectations, another child with the same genetic potential as the first one will come to behave very differently as he acquires different habits, skills, and beliefs prescribed by the second group.

GENES, CULTURE, AND BEHAVIOUR

The fact that sociobiologists understand that animals do not have within them a gene or genes ordering the individual to be automatically aggressive or to copulate once a day no matter what does not satisfy many social scientists. These students of human behaviour often argue that there is not even an indirect connection between human genes and human behaviour.

For Sahlins and many others, two aspects of human behaviour are taken as convincing evidence that human behaviour has not evolved to promote the survival of the genes of individual human beings. First, the critics not that if you were to ask any human why he did something, the person would never respond that he performed the action to elevate his inclusive fitness. But if a baby cuckoo could talk, it too surely would not tell an interrogator that it rolled its host's eggs out of the nest "because my genes made me do it." No animal need possess a conscious awareness of the ultimate reasons for its activities. It is enough that its proximate mechanisms predispose it to behave in ways that help its genes survice. On the proximate level we enjoy eating sweet foods, fall in love, want to be liked by our friends and relatives, and learn a language because we possess physiological mechanisms that make it easy for us to do these things. We do not have to be aware of the caloric content of honey and its contribution to cell growth and maintenance in order to eat it. It is sufficient that our brains tell us that honey tastes good. They evolutionary theory does not require that an animal know the genetic consequences of its acts, only that animals with certain abilities or preferences will tend to replace those with other aptitudes.

The second objection is much more challenging. Social anthropologists have documented that human cultures are amazingly diverse. To them this suggests that we are the completely flexible product of whatever social environment we choose to invent. How can it be said that humans tend to behave in ways that elevate their fitness when humans around the world behave in such different ways? These are polyandrous, polygynous, and monogamous societies; cultures in which females makes important political decisions and others in which males dominate the females. There are human groups for which warfare is a constant fact of life and other groups that never fight openly with one another. The list of cultural peculiarities is seemingly endless: depending on where you were born, you might be allowed to marry your first cousin or not; you might be allowed to look at your mother-in-iaw or not; you might be forced to have your penis cut from stem to stern at adolescence or not. Moreover, cultures are not only enormously variable, but they the capacity of change rapidly as seen in the transformation of human societies over the past 10,000 years from predominantly hunter-gatherer groups to the industrial high-density, high technology civilizations of today.

The indisputable diversity and flexibility of human behaviour potentially provide test after test of the evolutionary approach. We are an animal species. We do have witthin us genes. Therefore, evolutionary theory predicts that human behavioural traits, no matter how diverse of flexible, rest on evolved abilities that will tend to promote the genetic success of individuals. In fact, opponents of sociobiology could demolish not only sociobiology as applied to human beings but evolutionary theory in general if they could show that any basic human ability consistently reduces the inclusive fitness of individuals! Darwin recognized this general point, writing: "If it could be proved that any part of the structure of any one species had been formed for the exclusive good of another species, it would annihilate my theory, for such could not have been produced through natural selection". In modern terms, Darwin's challenge could be rephrased to read that current evolutionary theory would be annihilated if it could be shown that a trait originated exclusively for the genetic benefits of individuals that did not share genes with the practitioners of the trait. There are certainly some possible candidates for genetically altruisic human traits that by evolutionary terms should not have evolved if they really do result in lowered inclusive fitness: risking death in warfare, the use of birth control measures, and the practice of infanticide, for example, however, to show how these and other similar phenomena are all explicable by an evolutionary approach and that we can, at least for the moment, retain our confidence in evolutionary theory as applied to animals in general and humans in particular. But first let us consider why it is that cultural diversity may have evolved.

THE EVOLUTION OF CULTURES

Earlier chapters in this book presented evidence in support of two principles that are essential for an understanding of the possible evolutionary bases of human cultures. We begin with the realization that cultures are the product of the neural machinery within the human skull. Cultural practices are learned behaviours, transmitted from one generation to the next through tradition (teaching). One basic biological principle relevant to an understanding of cultures is that all learned abilities have a genetic foundations of some sort and, therefore, can and to evolve. In order for cultural practices to be acquired by each new generation, young human must have brains that are capable of storing the information contained in tradition. We can predict that to the extent culturally acquired information elevates individual fitness there will be selection in favour of humans who are able to absorb this information more completely, rapidly, or with less expenditure of effort than others. Conversely, if a cultural practice reduces inclusive fitness, individuals who were unable to learn it should enjoy a selective advantage.

Human brains appear to have properties that facilitate certain kinds of learning. The classic example is language learning: there is considerable evidence that the brain contains neural mechanisms that facilitate language acquisition and use. Simply because there is cultural variation in the languages spoken around the world is not proof of the independence of this behaviour from genetic influence. Genes are deeply involved in the production of the neural foundation of language, and it seems inescapable that alleles have been selected on the basis of how well "their" brains have been able to contribute to language use. If selection can produce humans with great skill at language, it should also be able to shape the design of the human brain so that humans can readily learn which individuals are relatives, how to use tools, how to behave in contain social settings, how to extract economic gains from the environment, and so on. As in the case of language, the successful genes need not produce neural mechanisms that are restrictively designed to permit only the learning of a specific cultural trait (such as the Turkestan of Jivaro or English language). Instead, genes may endow their brains with more open learning abilities, permitting a more flexible (but still not completely plastic) accommodation to whatever cultural variables dominate an individual's social environment. By this view, cultural evolution involves selection for various learning abilities that permit individuals to adopt that cultural practices of their societies. Thus genetic and cultural evolution complement one another rather than compete with each other.

But why have individuals with genes that indirectly helped them acquire a culture enjoyed a selective advantage in the history of our species? To answer this question it may be helpful to consider a second basic behavioural principle discussed earlier in the book: that social organization is a reflection of the ecology of the species. If a species confronts varied ecological conditions, it may evolve the ability to alter its social structure to match the conditions of the movement. If this is true for humans, culture may have evolved because the adoption of cultural practices enabled humans to invade a wide variety of environments each with its own special ecological pressures. Many authors have noted that cultural adaptation permits a more rapid adjustment to variable ecological circumstance than biological adaptation. If this hypothesis is correct, we should predict that cultural variation is not random; instead, the cultural practices of one group should permit individuals to exploit their environment more successfully than other cultural options. We should also expect that cultural changes elevate the inclusive fitness of the individuals that invent or adopt the change. People that practice the wrong option (one that results in lowered relative reproductive success) will be replaced by those who practice the superior option, transmitting this option culturally to their offspring. It seems obvious that the spread of tool use and the perfection of tool manufacture that occurred during human evolution were related to improve resource acquisition and the ability to produce and support more offspring. Moreover, the two major cultural innovations of the past 10,000 years—the invention of agriculture and the development of the use of fossil fuel as an energy soucre—have (1) spread extremely rapidly around the world and (2) been associated with a phenomenal increase in the human population. Both inventions, in effect have given participating individuals access to more calories and nutrients for conversion into offspring than non-participating individuals.

But what about those many peculiar cultural practices that actually appear to reduce the fitness of individuals rather than raise it? In some American culture it is customary to cook cornmeal in a 5 per cent solution of lime, pour off the solution, cool the mash, wash it, and finally use the prepared material to make tortillas. This involves a considerable expenditure of time and energy as well as the loss of some of the thiamine, niacin, nitrogen, and fat content of the corn. Although it would be easy to interrupt this method of corn preparation as maladaptive, S.H. Katz and his colleagues have shown that alkali cooking actually improves the nutritional quality of cornmeal (Table 7.1). It does so by breaking down the indigestible portions of the corn kernel and thereby freeing for human use substantial quantities of certain essential amino acids, including lysine, which is deficient in those parts of the corn seed that can be digested without this system of preparation. A diet that is dependent on cornmeal as a staple would be dangerous without alkali cooking; we lack the metabolic pathways to manufacture lysine, an amino acid that appears in many vital proteins. Moreover, alkali-treated corn has a lower ratio of the amino acids leucine to isoleucine, which improves its nutritional value. It is hard to imagine how someone invented this cooking procedure, but it is striking that all South and Central American cultures that rely of maize as the basic foodstuff have adopted the practice.

TABLE 7.1

Effect of Alkali Cooking on the Availability of Certain Essential Amino Acids

Amino Acid	*Milligrams of amino acid per gram of nitrogen in*	
	Uncooked corn	*Alkali-treated corn**
Histidine	0.012	0.028
Isoleucine	0.074	0.158
Leucine	0.309	0.358
Lysine	0.045	0.126
Threonine	0.145	0.381
Ratio: feucine/Isoleucine	4.2	2.3

*Com subjected to enzymatic degestion for 12 hours.
Source: Katz, Hediger and Valleroy [312].

Another cultural trait that has baffled outside observers is the sacred-cow tradition, which forbids or greatly restricts the sale or slaughter of cows for meat. In India the Hindu reverence for cows is such that they are allowed to roam freely, interfering

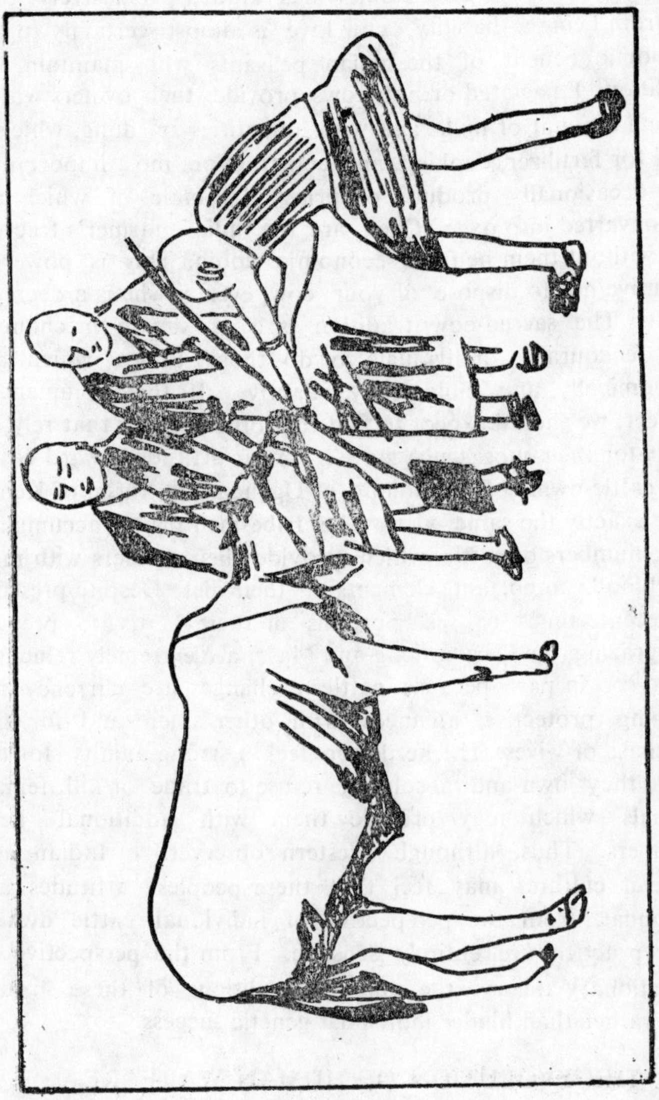

FIG. 7.1: Group of Masai Men.

with traffic, and they are almost never killed, even in times of famine. Why are not emaciated beasts sold or killed and eaten by their owners? Marvin Harris has argued persuasively that far from being a liability, cow love is almost certainly to the economic benefit of the Indian peasants who maintain the tradition. Emaciated or not, cows provide their owners with a certain amount of milk and with quantities of dung, which is used for fertilizer, cooking, and construction; most importantly they occasionally produce offspring, the males of which can be converted into oxen. Oxen are the Indian farmer's tractor, and without them he faces economic doom. This is powerful incentive not to dispose of your cow, even if she is a decrepit beast. The sacred-cow tradition is a proximate mechanism that encourages individuals to do those things which are economically and biologically adaptive. If this argument is correct, we should expect to find that other groups that rely on cattle for their subsistence have a similar attitude toward cows. The cattle-owning Karamojong of Uganda and Masai of Kenya have exactly the same view. The tribesmen try to accumulate large numbers of cattle, which provide their owners with milk and blood, important elements of their diet. Despite pressure in recent times to sell portions of their herds to prevent overgrazing, the Karamojong and Masai are extremely reluctant to do so, in part because cattle exchanges are currency for forming protective alliances with other men and for the purchase of wives. The herdsmen feel a strong affinity for the cattle they own and absolutely refuse to trade or kill female animals—which may provide them with additional herd members. Thus, although Western observers of Indian and African cultures may feel that these peoples' attitudes are irrational, from the perspective of individual cattle owners their practices are entirely sensible. From the perspective of evolutionary theory, the cultural traditions of these groups help rather than hinder individual genetic success.

THE EVOLUTION OF HUMAN WARFARE

Let us examine the complementarity of biology and culture in the context of one of the most dramatic of all human cultural activities, violent group aggression. The frequency and degree of aggressiveness exhibited by different cultures vary from the essentially pacifist societies of some Eskimos and Pygmies to the combativeness of the Yanomamo Venezuela, many New Guinean tribes in the recent past, and the societies of the United States and Russia, which are either constantly engaged in war or armed to the teeth in anticipation of the next round. This variation has been interpreted by some as evidence that war is largely a capricious feature of certain cultural environments and could easily be eradicated. Certainly the evidence indicates that the genotypes of humans do not blindly programme them to engage in warfare. But this does not mean that our genes have not endowed our nervous systems with special properties that provide humans with the capacity to engage in group aggression under certain conditions. From a evolutionary perspective it is possible that some ecological circumstances may so elevate the benefits of warfare to individuals that they exceed the costs to individual inclusive fitness. An argument that the abilities that underlie the capacity for warfare are generally advantageous does not imply either that warfare is inevitable or that it is morally desirable. This does not prevent me from accepting as a hypothesis that the traits that may be employed in group aggression can help maximize the inclusive fitness of the humans that possess these traits. Here we shall examine the plausibility of the hypothesis, taking this example to illustrate how (1) evidence about our history as a species, (2) comparative behavioural data from other living animal species, and (3) ecological comparisons of human cultures may be brought to bear upon the question.

The Evidence from Human History

Group warfare is an organized and usually premeditated aggression with the goal of destroying or maiming other human beings whom the aggressors may not even know. Perhaps paradoxically, it is promoted by the cooperative abilities of members of each warring group. The great capacity or co-operative among the members of human clans, tribes, villages, and nations can be seen as the product of powerful selection forces operating on our ancestors. Although experts disagree on some details of the interpretation of the fossil record, there is nearly universal acceptance of the following points. Our hominid forebears were group-living primates, closely related to the great ape lineage, whose key ecological invention was to become a hunter of living animals in addition to gathering vegetable foods. The use of tools became a central adaptation that enabled protohominids to capture and process game efficiently. The importance of this culturally transmitted

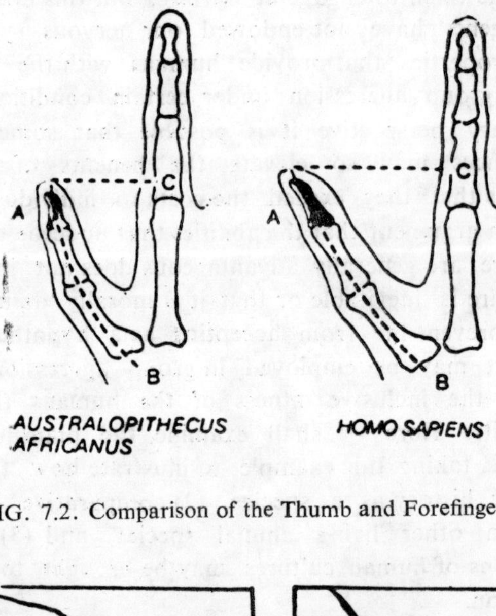

FIG. 7.2. Comparison of the Thumb and Forefinger.

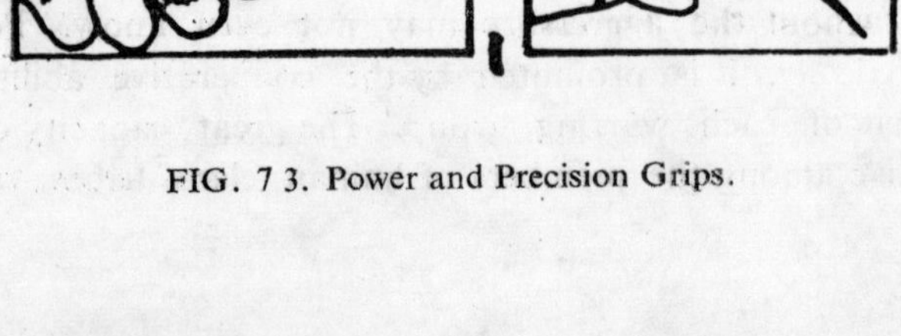

FIG. 7 3. Power and Precision Grips.

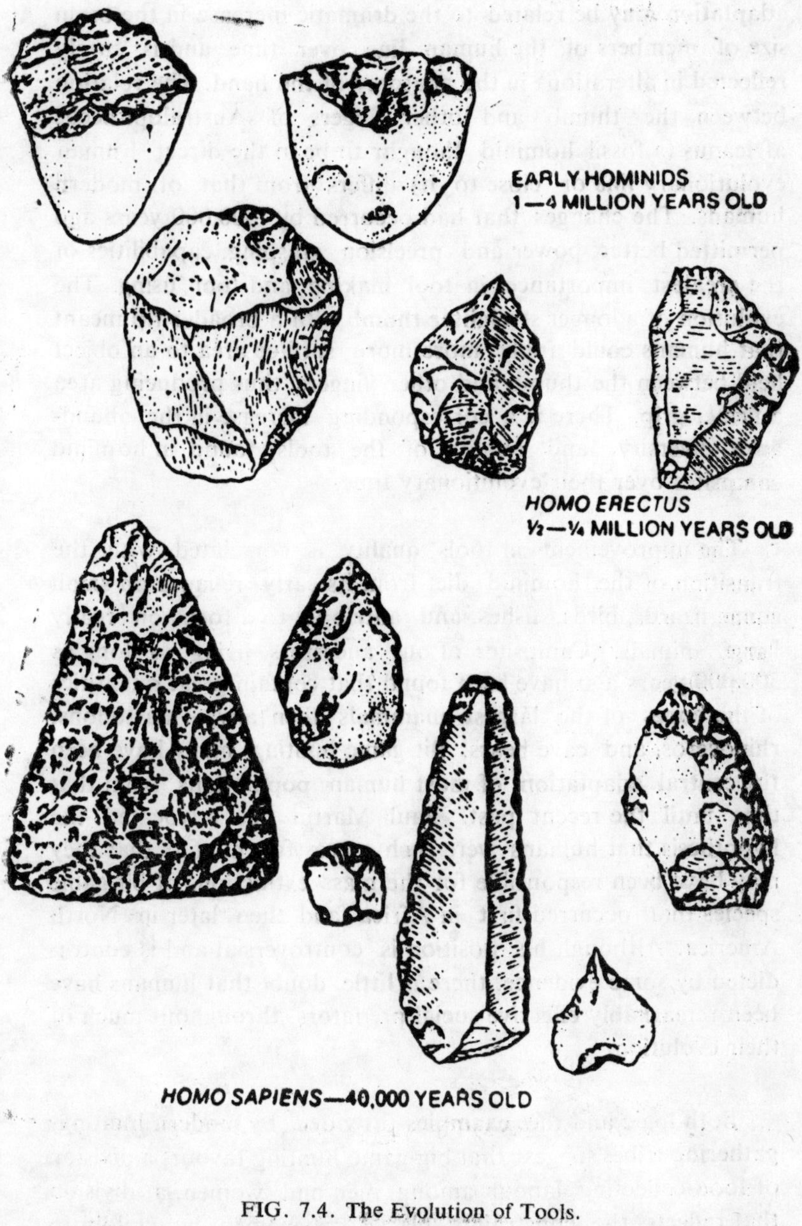

FIG. 7.4. The Evolution of Tools.

adaptation may be related to the dramatic increase in the brain size of members of the human line over time and is surely reflected in alterations in the anatomy of the hand. The relation between the thumb and other fingers of Australopithecus africanus (a fossil hominid thoughr to be in the direct human evolutionary line or close to it) differs from that of modern humans. The changes that had occurred by 500,000 years ago permitted better power and precision grasping, capabilities of the greatest importance in tool making and tool using. The evolution of a longer straighter thumb with a broader tip meant that humans could apply more more surface area to an object held between the thumb and other fingers, thus producing area a firmer grip. There is a corresponding increase in the abundance diversity, and quality of the tools found in hominid campsites over their evolutionary time.

The improvement in tools quality is correlated with the transition of the hominid diet from its early reliance on small game, lizards, birds, fishes, and antelopes to a focus on really large animals. Campsites of our ancestors living as long as 300,000 years ago have been found that contain great quantities of the bones of the largest mammals then alive—mastodons, rhinoceros, and cave bears. Bit-game hunting many have been the central adaptation of most human populations from that time until the recent past. Paul Martin has advocated the hypothesis that humans were such successful hunters that they may have been responsible for the mass extinctions of bit-game species that occurred first in Africa and then later in North America. Although his position is controversial and is contradicted by some evidence, there is little doubt that humans have been remarkably effective social predators throughout much of their evolution.

Both logic and the examples provided by modern hunting-gathering tribes suggest that big-game hunting favours a division of food-collecting labour among men and women, a division that reflects the constraints placed on women by childbirth and infant care. Selection pressure on human females has produced traits that facilitate bearing and rearing children. But a wide pelvis that makes the birth of a large-brained infant possible also makes locomotion silghtly less efficient and, therefore, interferes with hunting (just as selection exerted on male lions to be effective territory defenders has favoured the evolution of a large conspicuous mane, which makes prey capture more difficult for miles). Dividing the gathering of food resources between the sexes permits specialization and thus presumably greater efficiency in exploiting a diversity of foods.

Hunting large and sometimes dangerous animals should favour individuals capable of cooperation in planning and executing a successful hunt. Moreover, the savannah habitat in which large herbivores are found and which our hominid ancestors invaded in search of game also is home to potential predators of man. The danger of isolation and the advantages of group defense may have been major selective agents promoting tight social cohesion among the members of a band. In addition, the cooperative abilities useful in hunting big game and repelling predators could also be employed to kill or drive off other competitive species—the large carnivores that competed with humans for the prey herbivores of the savannah. Finally, the great likelihood that members of a band were probably close relatives would have further elevated the benefits of intraband cooperation. An individual capable of working with others for a common goal would not only be improving his or chances of reproducing, but might also be helping increase the fitness of other humans with many shared genes.

The Evidence from Others Animals

The plausibility of this sketch of the evolutionary events and selective forces that set the stage for human warfare can be examined by making some careful comparisons of human behaviour with that of other animals. A common complaint of critics of an evolutionary analysis of human behaviour is that sociobiologists make caplicious use of comparative meterial, selecting whatever species help support the hypotheses favoured by sociobiologist. However, as has been outlined in previous chapters, there are species' to gain insight into the ecological and historical bases of a behavioural characteristic of interest. First, we can look at closely related species to determine which of their behaviour are similar to our own. This information can help us determine the history of a behaviour pattern. Second, we can examine unrelated species that experience similar ecological pressures to determine if they have evolved convergent behavioural solutions to these shared problems. For humans, therefore, the species whose behaviour may be most useful for an understanding of our own abilities include (1) other primates, especially out closest relatives, the chimpanzee, gorilla, orangutan, and gibbon, and (2) the mammalian social cornivores, which are not closely related but which face the same obstacles associated with capturing big game.

Human Sex Roles in Society

The relative roles of the hereditary and environmental components in the expression of human behaviour has been a matter of heated debate in relation to the topic of human sex roles in society. Many persons expect a female to behave in a certain manner and a male to respond to a particular situation in a characteristic way which is distinct from that of a female. Generally, the female is expected to be more retiring, the male more aggressive. For a male or female to behave in a fashion considered inappropriate to his or her sex is often labeled unnatural. According to the ideas of some students of behaviour, characteristic male and female behaviours are fixed in the species. There is a genetic basis, according to this thesis, which is responsible for the expression of male social behaviour and of female social behaviour. The argument of the proponents of this idea is based mainly on studies of the fossil record, observations made on the behaviour or various species of primates (the order to which the human belongs), and observations of behaviour patterns in variou human societies today. Proponents of the concept attribute the fixation of sex roles to the advantage this behaviour imparted to the human species during its evolution in the relatively recent past. Social behaviour in which sex roles are distinct would have given a decided benefit to the human when he become a hunter. Once the human species assumed hunting activities, so the argument goes, an assignment of roles to each sex would have given insurance to the success of these activities and hence would have increased the survival of the group. Any genes or genetic combinations which would magnify the differences in sex role behaviour would be selected under the force of natural selection. As the hunter, the human male would require strength, endurance, speed, alertness, and prowess among other attributes. Such hunting efforts would benefit from improvement of these traits, and so any genetic determinants in this direction would be selected. However, such traits would benefit directly only the active hunter, the male. Interactions of genes for excellence in these traits with the hormonal environment would fix them in the male. On the other hand, any hormonal controls to make the females more different from the male would have value. The changes in the female during this part of human evolution would have occurred only in relations to the requirements of the male in his role as the hunter. In her role as childbearer and rearer, intelligence would not be at a premium. Cunning, strength, and swiftness would not be needed for food gathering activities. Genetic controls to make the female subordinate and docile would benefit the needs of the male after the demands of the hunt and would in turn make him a more efficient hunter. Hormonal regulations in the female to make her sexually recepitive at any time would have value. They would have been selected in relation to the need of the hunter who would be assured that the female would be available to him whenever, he so desired after return from the hunt. Strong bonds would form between males. Such bonding would be of advantage since it would assure cooperation during hunting activities and would reduce hostility among the males. No souch bounding among females would be required, since the female's activities did not depend on major group activities. Therefore, sex roles became more and more distinct as the success of the hunt became more and more a certainty.

According to this idea, we see today in our human societies behavioural traits which characterize the male and those which are typically female. These exist because they are set in the genetic programme where they became established by natural selection during the time the human emerged as a hunter. So today, mainly the men perform decision-making activities. Women accept these decisions. The male is aggressive, whereas the females is the homebody concerned primarily with child rearing and pleasing the male. All of the behaviour patterns which we today consider male or female are largely a direct result of the genetic factors which control these activities and which became established in the past in relation to the hunting activities of Homo sapiens.

There are, of course, many arguments which can be offered to counter these ideas. Not the least is the demonstration that human behaviour is highly plastic, a feature which distinguishes it from that of other species. There is good evidence to show that an individual, regardless of sex, may be just as intelligent, docile, aggressive, or strong as any other, depending on the experiences of the individual during development. No evidence exists for some sort of maleness or femaleness in behaviour which is set and cannot be modified. There is no clear-cut evidence from the fossil record to reveal details of the social life of the human when the species emerged as a hunting society. It is possible that some of the participants in the hunt were actually females. The studies of behaviour in other primate groups should be viewed with suspicion when conclusions are drawn from them and applied to the human. First of all, one should remember that the human is a human and the monkey today is a monkey. To draw conclusions on the basis of an animal's behaviour today and to apply it to humans who once lived in the past are misleading tactics. All of our biological evidence indicates that no animals as it exists today is ancestral to any other which exists today. While they may have had a common ancestor in the distant past, this ancestor may have behaved very differently from either one of the modern groups today. In short, today's human did not arise from today's monkey or ape. Moreover, conclusions are often drawn from groups in which the male happens to be dominant. It some primate groups such as the gorilla, there are no sharp distinctions which set off the male as the strictly dominant sex. In addition, many conclusions are drawn on the behaviour of animals in captivity, a factor known to distort an animal's behaviour. Studies on so-called primitive human societies today provide no concrete evidence to the genetic assignment of sex roles. The primitive populations of the human species cannot simply be assumed today to be exactly like those populations in existence millions of years ago. We don't known that much about the history of these groups today nor of the cultural changes in them which are very recent adaptations.

While a genetic component certainly exists for species-specific behaviour, we must recognize the advantage the human has in the flexibility which his gynetic endowment permits in his behaviour. It would seem less wise, on the basis of highly debatable arguments. To assert that certain types of behaviour are programmed or inherent than to appreciate the range of human responses made possible by the very plasticity which the genetic endowment allow. A fuller understanding of the subtle interplay between genetic factors and environmental influences in the establishment of human behaviour can lead to a deeper appreciation of differences among groups and individuals. It may also offer clues to the alleviation of serious problems which can arise from certain forms of human behaviour.

THE EVOLUTION OF HUMAN REPRODUCTIVE BEHAVIOUR

Warfare is behaviourally highly complex, and its relation to inclusive fitness not immediately clear. Reproductive behaviour is much more obviously linked to individual fitness, and, therefore, the imprint of selection should be readily apparent on it. Yet is is well known that there is considerable variation within and among cultures with respect to many reproductive practices. Moreover, there are actions, including mate sharing, adoption of another's child, and the practice of birth control, which, on the face of it, appear to reduce fitness. In this section, we shall look carefully at some aspects of cultural diversity in reproduction and cases of appearent genetic altruism to see whether they refute the evolutionary prediction that humans will behave in ways that tend to maximize their inclusive fitness.

The family unit is universally accepted by anthropologists and others as the basic component of human societies. Human males and females have the capacity, shared with many other animals, to form a family in which male and female live together and cooperate in the rearing of the progeny they produce. The selection of a mate should have a great deal to say about the genetic success of individuals. What evidence is there that human males and females behave in ways that promote their genetic success in this phase of the reproductive process?

Mate Selection

One significant feature of mate-selection is the culturally reinforced prohibition against copulation between some sexually mature individuals. Incest taboos appear to be a universal feature of human social behaviour. Thus parent and their children are forbidden to copulate, and people are usually not permitted to marry their siblings (and often first cousins). Incest taboos are often culturally extended to forbid marriage between individuals belonging to particular categories, such as the same clan. The strength of the incest taboo has been demonstrated by the refusal of children reared together in Israeli kibbutzim to marry one another. When the kibbutz system was devised, the social scientists in charge envisioned that communally raised children would eventually marry within the kibbutz. Such marriages were actually encouraged, and yet not a single one has occurred; the children of one kibbutz always fall in love with members of another kibbutz. It is as if the brain had a programme that reads: "Do not become sexually attracted to humans with whom you have been reared." (When unrelated cactus mice were taken from different litters and experimentally reared together, they too were reluctant to copulate with nestmates. Over evolutionary time juvenile mice, and humans, that grew up together were almost always siblings or very close relatives). The inability to become sexually attracted by close relatives contrasts sharply with the ability of humans to fall in love with complete strangers. This ability, the incest taboo, and the culturally extended proscriptions against copulating with or marrying certain classes of individuals all contribute to outbreeding. The occurrence of similar mechanisms in other animal species suggests that a major advantage of these traits is the avoidance of inbreeding. This prevents the accumulation of deleterious recessive alleles in double close in one's progeny. Humans from incestuous matings exhibit a substantially higher rate of genetic defects and congenital abnormalities than members of outbred populations. Thus humans should benefit genetically to the extent that they avoid inbreeding. An added benefit to outbreeding among humans is the formation of alliances between different genetic lineages. The progeny produced in a marriage between two separate family groups creates a tie between the two bands that may promote cooperation between them. Without a mutual interest in the children of the marriage, the two groups might have no other shared concerns and no motivation to help and another.

Female Choice

The incest taboo applies to both males and females, but there is some indication that male parents break the taboo (by copulating with their daughters) more often than females. This may reflect the lower threshold for sexual excitation among males that females that is characteristic of animal species in which males may pass on their genes at little expense relative to the female. Female humans, like all female mammals, must make a large investment in each progeny while it is in the embroyonic stage and during the early years of its life when it is nourished at the breast. Because a male's investment in an offspring may be as little as the sperm used to fertilize the egg, evolutionary theory predicts that females should be more selective than males in the choice of copulatory partners. There is ample anecdotal evidence as well as scientific data collected by Masters and Johnson that females of our culture are less easily sexually aroused than males. Females should be more cautious because they have much more to loss than a male. If a woman becomes pregnant at an inopportune time (*e.g.*, when she is unmarried and her partner has no intention of helping her rear her child), her large physiological investment may be poorly rewarded. At least in the past, the child of a husbandless woman would have had a reduced chance of surviving to maturity; if it did reach reproductive age, it would probably have a lower fitness than average because of juvenile malnourishment or the possession of few material goods with which to attract mates. Moreover, a single mother may reduce her long-term chances to acquire a helpful mate, either because potential husbands would consider her a poor risk or because they are uneager to help care for another man's offspring.

Thus in cultures in which males may provide material resources for their children, a female should permit copulation to occur only if her partner has demonstrated both a willingness and an ability to make a large parental investment in their progeny.

Male Parental Investment

It is easy to see why females might prefer mates willing and able to provide them with food and protection. If a division of labour between the sexes has been the rule throughout much of human evolution, males have controlled highly valuable resources (such as protein-rich game in many societies), which when used by a female and her progeny would greatly enhance their fitness. But, as discussed earlier, the male animal typically enjoys reduced benefits relative to the female from equal parental investments per offspring. First, the reliability of paternity is often less than the reliability of maternity, and a male therefore runs the risk of helping someone else's genes. Second, the cheapness of sperm means that males can potentially inseminate many females. This elevates the benefits of a reproductive strategy that emphasizes effort to copulate with many females rather than invest heavily in one or a few mates. Male mammals, including most primates, rarely are monogamous and rarely provide much in the way of parental care for the progeny of females with which they have copulated. Why are human males an exception to this rule?

The following explanation is speculative. We begin with the hypothesis that males have consistently been more successful hunters than females. A female protohominid in estrus might attract a consort male who would guard her during the time she was receptive and most likely to ovulate. During the consort phase a male might share prey he had recently captured with the female because it would be likely that the progeny produced by the female would bear his genes; protein gifts to the female might, therefore, have genetic gain for the male. (As usual there is no requirement that individuals have to be aware of the genetic consequences of their acts). If gift giving became an established component of courtship and consort guarding, than a female might gain if she could extend the period of guarding and thus receive more presents. Instead of blatantly advertising the time of ovulation by becoming receptive just prior to this event, female protohominids might have partly hidden the time of ovulation by lengthening the period of receptivity. The adverisement-of-ovulation strategy is practiced by female mammals as an incitation display to encourage competition among males for access to them, thus increasing the probability of mating with a dominant male. But if parental investment is offered by a male, the quality of this investment may become more important to a female than signals of male genetic quality. By making if difficult for a male to determine when his chances of fertilizing an egg were highest, females that retained their receptivity could make it advantageous for at least some males to practice prolonged guarding. The more hidden the ovulation time and the more prolonged the period of receptivity, the longer a male must remain with a female (copulating with her at regular intervals) if he is to increase his chances of fertilizing her eggs when they do become available for fertilization. A widespread use of a strategy of permanent guarding (marriage) reduces the availability of fertilizable females and reduces the time a married male has available to find these females (because of the guarding time required to protect a mate). This being so, it may become advantageous to be a devoted husband and father and to expend resources on one's presumptive offspring, whice are highly likely to bear one's genes. This may increase their chances of surviving and reproducing, providing a genetic gain that may often (but not always) exceed a Casanova strategy of attempts at multiple sexual conquests.

Because human males can offer useful parental services to their offspring, one can expect females to choose husbands partly on their potential ability to provide superior investments (Table 7.2), whereas males should make the parental investment in ways that tend to promote their own genetic success. A female should choose a husband who apprears to be genuinely devoted to her and whose gifts and reputation are evidence that he is willing to help her. Males can and do deceive some females, seducing them with the intent of deserting them at some point. Female coyness and caution about copulating have the function of reducing the probability of making mistakes about males.

Women should also be concerned about the economic status of males, present and potential. The fact that females in a great many human societies willingly enter into polygynous marriages is at first glance puzzling. One might predict that given the potential role of males in child rearing, females should prefer to male with a single male. Carnivorous mammals generally show a greater tendency toward monogamy than herbivorous ones. In fact, the vast majority of human marriages have surely been monogamous. Nevertheless, why should there be any polygynous marriages freely entered into, given the males potential contribution to the welfare of his wife and children? Perhaps human men are rather like redwinged blackbird males. Most human societies offer opportunities for truly exceptional performance in various fields, with commensurate material rewards. Outstanding leaders, hunters, weapon makers, and healers have probably always secured large amounts of food directly or indirectly. Because the difference between successful and unsuccessful men is fairly obvious, marriageable women have probably had no difficulty discriminating between the two. Given a choice between an outstanding provider and an inferior one, it may be to the women's advantage to choose the former, even though this means entering into a polygamous relationship. In general, the economic status of a potential husband should be at least as important as his physical attractiveness (and those male traits that are generally considered attractive—a well-proportioned body, broad shoulders, tallness, strength, and intelligence—are the kinds of attributes useful in hunting and in competing with other males both within and outside the group).

Male Choice

As should be apparent from the preceding section, the reproductive interests of males and females are not necessarily convergent and may often be at odds. I believe this possibility is related to the low threshold for suspicion and fear of exploitation that characterizes many interactions between the sexes (and which is well developed in the feminist movement and in the hostile response of many males to militant feminism). Males, like females, appear to behave in ways that promote their own special reproductive goals. Because of the costliness of male parental care, we can expect human males to be selective about helping children. They are. Because the reliability of paternity can never equal the reliability of maternity in our species, husbands should be somewhat more cautious than their wives about providing material resources for their children. We predict the willingness of a male to remain with his wife and help her children should be proportional to his certainty of paternity.

TABLE 7.2

Characteristics of Males that are of Potential Interest to Females in Selecting Mates

Degree of relatedness
Fathers, brothers, proscribed relatives not acceptable mates
Willingness to provide parental investment
Quality of attention given female by suitor
Gifts given to female by suitor
Reputation of suitor
Ability to provide parental investment
Quality of gifts given to female by suitor
Resources known to be controlled by male
Dominance status of male within the society
Intelligence

Suspicion of adultery and fear of cuckoldry are extremely widespread phenomena in cultures in which males take care of their children. These emotions often play a major role in the social relations of husbands and wives. The ultimate cause for male concern about these matters is the severity of the genetic consequences of female adultery for a providing husband. If the child he helps lacks his genes, he is not only failing to promote his reproductive success but actually elevationg the fitness of a competitor. From a female's genetic viewpoint, a child, whether the product of an adulterous liason or not, will have 50 per cent of her genes. In many societies, male and female adulterers receive harsh treatment at the hands of aggrieved husbands. Among the relatively peaceful Bushmen of the Kalahari Desert, most murders are committed by men in disputes over women, especially if adultery is involved.

Even if the reduction in certainty of paternity is not the female's fault, as in the case of rape, a husband may still divorce his wife, a response accepted or even encouraged by a variety of religious groups and cultures. Raped women often do not report the incident to their husbands for fear of the reaction.

The protectiveness of husbands about their wives and their reluctance to continue to care for adulterous (or raped) wives and their progeny seem to confirm evolutionary predictions about how males should behave to maximize their inclusive fitness. But what about wife sharing, a practice not unknown in our society and fully sanctioned in certain other cultures? A speculative interpretation of wife sharing is that it may be an extreme example of reciprocal altruism. If the sharing is truely equal, each husband stands the same chance as the other of eventually providing support for offspring fathered by the other individual. Therefore, neither male necessarily reduces his fitness if he sees to it that his generosity is fully reciprocated. Moreover, if the exchange of wives helps cement an alliance between the males that enables them to cooperate more effectively in the acquisition of resources, the fitness of both males might be elevated in the long run. Male that exchange the right to copulate with their mates do not lack the emotion of jealousy or the sense of possessiveness about their wives, as is shown by the violent treatment or death that adulterous (non-reciprocating) males receive in some Eskimo groups and other tribes in which wife sharing.

Any number of cultures are far more sexually permissive than our own, and in them adulterous sexual relationships may be commonplace. The sexually open South Sea Island societies of the Pacific are a classic case. In numan groups of this sort, the reliability of paternity may be very low. The standard practice within such cultures is for males to withhold parental care from the children of their wives and instead to adopt the father role for the children of their sisters. In our society we consider it natural for mothers and fathers to love and care for their own offspring. There are many proximate mechanisms that encourage individuals to do this. For example, it is now recognized that a mother usually forms a strong, loving attachment to her baby in the first few hours after it is born. Incidentally, the recent hospital practice of separating the neonatal infant from the mother immediately after it is born and keeping contact to a minimum for the first few days of its life almost certainly interferes with the development of the bond. If a strong bond is not established at this time, this may have permanent negative effects on the relationship of the mother to her child. More generally, parents, male and female, are disposed to find their children truly delightful and charming creatures, even when more dispassionate observers are hard pressed to see why. This is especially true of helpless newborns and younger children, who possess certain facial features (*e.g.*, chubby cheeks and large eyes) that are exceptionally attractive to adults.

From an evolutionary perspective, parents may offer their offspring love and care for many years because these individuals usually have a high proportion of their genes. But a male whose wife's children were fathered by other males does not share genes in common with these offspring. If most or all of his wife's children are genetic strangers, it may become adaptive to share resources with his relatives instead of his wife and her progeny Richard, Allexander has shown that if, on the average, a male fathers only one of four children produced by his wife, he would gain genetically by diverting his parental care to the children of his sister. Under these conditions, a male shares 5/32 of his genes on the average with his nephews and nieces, but only 4/32 with his wife's children. Importantly in this society three times out of four a male helping his wife's offspring will be assisting an individual with no shared genes at all. In contrast, by becoming an uncle-father, he will he helping a sister's progeny. All of whom have at least some of his genes by common descent.

Because humans everywhere display an enormous interest in and aptitude for learning "who is related to whom and by how much," they are in a position to regulate the distribution of their resources and services in ways that could help them maximize their inclusive fitness. The ability to assume the uncle-provider role in certain circumstances is merely one example of the genetic utility of knowning who is related to you and being predisposed to provide parentlike assistance in accordance with the degree of relatedness.

Adoption

Two aspects of human reproductive behaviour provide what may be the most difficult challenges to an evolutionary approach to over behaviour. They are the practices of adoption and birth control. We all know people in our society who have adopted into their families the child of a complete stranger, which they treat with the same affection and generosity that is customarily offered to one's biological children. These adoptive parents are, in effect, subsidizing competitor genes. How is it possible to account for such a maladaptive response in evolutionary terms?

Advocates of an evolutionary approach tackle problems of this sort in a special way. If the action is highly unlikely to increase inclusive fitness, as is the case for the adoption of a complete stranger, there are two alternative evolutionary hypotheses that may help explain the behaviour:

1. The action is one of the costly effects of an ability whose benefits, on average, exceed the costs.
2. The action is currently disadvantageous solely because of the radically altered setting in which it is now performed; in the past, the disadvantageous response would not have occurred.

These explanation for traits that seem to harm an individual's chances for genetic success have been used throughout this text. For example, the copulatory response of a male bee to an orchid flower or the flight of a moth to a bola spider were held to be disadvantageous aspects of the generally adaptive ability of these individuals to find receptive females of their species. A hypothesis of the second type was employed to explain why robins sometimes hurl their banded offspring from the nest. The silvery band stimulates the nest-cleaning response, which is used in a way that would never have occurred in the past because human bird banders are a recent, novel addition to the bird's environment.

Similar hypotheses may help explain the modern practice of the adoption of strangers. First, adopting a nonrelative may be a genetic cost of the usually adaptive desire of adult humans to have children and raise a family. Although it is true that adults who adopt strangers are reducing their fitness, the urge to have a family and the love of children that cause them to behave this way are on average beneficial. Because them traits usually elevate the fitness of the humans that exhibit them, they are therefore maintained within human populations. The point is that all adaptations have costs and benefits, and one cost is that they may be misused, from a genetic perspective, especially in rare or unusual circumstances. In Joel Welty's The Life of Birds there is a photograph of a cardinal feeding a beakful of insects to a goldfish that has lunged out of the water, mouth gaping, to receive the food. This act certainly could not benefit the cardinal's genes. The bird, however, had recently lost its brood and was employing its food gathering and brood feeding in a rare and inappropriate manner. The ability to collect food for young and the strong desire to feed them surely elevates the fitness of most parent cardinals.

The misplaced-parental-care hypothesis generates a testable prediction, which is that husbands and wives who fail to produce children themselves should be especially prone to adopt strangers. Their frustrated desire to be parents could find inappropriate expression (from an evolutionary perspective) in the adoption of a biological child substitute, such as an animal pet or an unrelated infant. However, the risk that parental drive will be misused is relatively slight, and the overall genetic benefit of the eagerness to have children is obvious.

Another prediction is that if the adopting parents proceed to have children of their own (not an uncommon occurrence), then they will be more likely to withhold care and resources from an adopted child of strangers than from their own biological children. One way to test this prediction would be examine the victims of child abuse in a family with both natural and adopted children.

The second basic alternative hypothesis is that the adoption of strangers is a novel, recent phenomenon that occurs primarily because of changes in the typical human environment over the past several hundred years. In large industrial societies, children are made available to people who do not know the parents of the adoptee. Throughout the vast majority of human evolution, in which people lived in small bands or villages, this event would be exceptionally rare. Adoption in this environment may have been practiced regularly to the genetic benefit of the adopter. There are at least three possible kinds of adoptions that could raise the inclusive fitness of the adopting individual.

1. Almost certainly acceptance of the children of relatives was the most common kind of adaption in the past. A direct way to elevate the fitness of a relative, and thereby perhaps to raise one's own inclusive fitness, would be to assist the children of that relative. In effect, males that practice the uncle-father strategy are adopting their sister's children as their own. Adoptive assistance given related persons might also occur frequently when the biological parents died or experienced some disastrous turn of fortune that greatly reduced their ability to care for their children. By stepping in to help, an individual could salvage nieces and nephews whose reproductive potential would otherwise be nil or very low.

2. In cases in which an individual adopted a stranger, the adoptee might be a captured child from an enemy group. The adopting family would probably not treat the child in an altruistic fashion, but instead exploit him or her as a servant or slave.

3. There are some well-documented cases of self-sacrificing adoption in preindustrial societies of individuals known not to be relatives of the adopters. Marshall Sahlins, who believes this phenomenon cannot be explained by recourse to evolutionary theory, also points out that "adoption is, like marriage, a mode of alliance between group", thereby providing a possible evolutionary explanation for adoption of this sort. A reciprocal altruism hypothesis predicts that for adoption of nonkin, the adopting individuals, although temporarily lowering their fitness by adoption of a non-relative, in the long run elevate their fitness by establishing a mode of alliance with the parents (and kin) of the adopted child. In the Pacific Island cultures where this practice was widespread, warfare was also highly developed. Adoption among non-relatives may have increased the possibility of alliances between villages for mutual benefit in defensive or offensive war maneuvers. Such alliances are often critical in determining the survival of a group in cultures where warfare is a major activity. Adoptive practices may also have decreased the probability that allies would turn on each other during or after the war because they would have a direct genetic stake in the survival of the other group. Finally, if one could adopt the child of an enemy tribe, one might reduced the likelihood of continuing aggression between the opponents and so end or moderate hostilities. In this context it is significant that the warlike Tahitians were willing to adopt the child of a person they had killed in a battle or a child of a close relative of the deceased person.

BIRTH CONTROL

How is it possible to reconcile the contention that people love children so much so that they will sometimes voluntarily adopt the infant of a stranger with the widespread current and ancient practices of birth control, including abortion and infanticide, all of which prevent people from having children? An evolutionary approach to this problem begins with the understanding that one can sometimes maximize fitness by not reproducing under certain conditions. The reproductive physiology of humans reflects this principle. Rose Frisch has shown that food intake helps control both the onset of menstruation in women and their fertility during the childbearing years. In fact, a minimum of 16 kilograms of stored body fat is required for maintenance of menstrual cycling. This reserve contains 144,000 calories, an amount than sustain the development of a typical fetus and 3 months of lactation for the infant after its birth. A sharp weight loss (or emotional stress) can cause the physiological blockage of ovulation and, therefore, the prevention of pregnancy. The mechanisms, pormonal and otherwise, that make this happen are performing unconscious birth control for the person, preventing the formation of azygote that has a poor chance of surviving and a high probability of simply draining resources from a mother with no genetic return.

Conscious behavioural decisions about when to produce children can also have the effect of raising rather than lowering fitness by increasing the total number of reproducing offspring created by a parent during his or her entire lifetime. It is possible to make some predictions about when humans in a wide variety of cultures will practice birth control, if the ultimate function of this behaviour is to maximize individual genecopying success.

Prediction 1. Unmarried women in societies in which the biological father normally assists in child rearing should be more likely to practice birth control, including abortion and infanticide, than married women, for reasons already discussed.

Prediction 2. Married couples will practice birth control primarily with the goal of spacing the births of their children at intervals of several years. In many cultures, a made is forebidden to have sexual intercourse with his wife for months after the birth of a child. As you might suspect, this reduces the probability that the female will become pregnant during this time, an event that would divert metabolic resources from milk production to embryonic development. Such a diversion would often endanger the survival of the living infant. If a newborn can survive the first few years of life, it stands a better and better chance of surviving to reproduce and, therefore, becomes increasingly valuable and deserving (in genetic terms) of continued investment. In hunter-gatherer cultures, nursing continues for up to 5 years after the birth of a child. Nursing activates hormonal systems that help prevent ovulation; cultural proscriptions against copulation between new parents reinforces the ultimate effect of the antiovulation mechanisms, which is to space births widely.

There is some peripheral evidence that suggests that humans are designed to have children at intervals of 3 to 5 years. In modern cultures, some people are able to have many children with very short periods between each birth. Psychologists have found that the intelligence of children (which may influence their potential reproductive success and thus the fitness of the parents) is significantly influenced by both birth interval and family size. Later children in large families have considerably lower intelligence scores on the average than earlier-born progeny. Likewise, the longer the interval between births (up to 4 or 5 years), the higher the scores of children on intelligence tests. These results may indicate that human developmental systems are designed to operate in a social environment with siblings 4 years or so older or younger or that parents may be better able to assist in the intellectual development of their children if they are widely spaced.

Despite hormonal and behavioural adaptations designed to keep a mother with a very young child from becoming pregnant again, it may still happen. Under these circumstances, aborting the fetus or killing the newborn may in evolutionary terms represent ending a geneticallyunwise investment in reoroduction earlier rather than later, thus saving energy to devote to living offspring and to promote the long-term reproductive success of the parent.

Prediction 3. In cases in which infanticide does not have the effect of spacing attempts to rear offspring, it will result in the removal of progeny whose potential for genetic success is relatively low. Thus, we can predict infanticide to occur with higher-than-average frequency when the newborn has a congenital defect apparent at birth or when twins are born. (Because of their low birth weight and because their mother's milk must be shared between them, twins run a higher risk of mortality than single children).

It is possible to interpret the selective infanticide of healthy newborn females in some cultures in the light of this hypothesis. Female infanticide is practiced primarily in highly warlike cultures in which male mortality produces a skewed adult sex ratio. In the absence of female infanticide, the sex ratio would become even more strongly unbalanced in favour of females. Because such societies usually exhibit a polygynous mating system, a surviving male child in a population with many females may have a far higher potential for genetic success than a female whose reproductive success is more certain but limited by the relatively small number of children she can bear. This favours couples who produce male infants. If a mother rears a female newborn, there will be a long interval (3 to 6 years) before she can expect to conceive again. By killing her female infant, a woman can quickly become pregnant again and perhaps this time produce a genetically more valuable male child.

Prediction 4. Life-long celibacy will almost never be practiced by children that are the sole offspring of their parents. Individuals who never engage in sexual intercourse are rare, but do occur regularly in human societies. Females in this category "should" devote themselves to their relatives' children, preferably their nieces and nephews who share a substantial proportion of genes in common with them. The tradition of the old maid aunt is well established in our society and others. Celibate males are usually found in the priesthood. We might predict them to occur more frequently in polygynous cultures than monogamous ones, with the celibate role being adopted by males who could judge (probably unconsciously) that their chances of competing with other males for mates were poor. By proclaiming their rejection of personal reproduction, they might be better tolerated by reproductively active males (perhaps sometimes giving them a chance to deceive these males). In the role of religious leaders, they might also be in the position to donate goods received for their services from non-relatives to their relatives, but only if they had relatives to help. Thus a male, if he were an only child, would generally have less to gain genetically by joining a celibate priesthood than if he were to reproduce personally. A male from a large family might have more opportunities to elevate his inclusive fitness as a celibate priest than by trying to form a family of his own.

We have argued that birth control can be and often has been used in ways that contribute to individual genetic success. It is indisputable, however, that in some modern societies birth control is used by many couples to reduce their reproductive potential. To explain this misapplication of birth control (from the gene's standpoint), it may be sufficient to note that reliable means of preventing pregancy (other than abstinence which has never been popular) have become available only in the last few years. This provides couples with the unique freedom to copulate without the subsequent burdens and demands of parenthood. It is striking, however, that almost all husbands and wives in western society, even with access to nearly perfect birth control measures, still choose to have several children. However, they tend to draw the line at two or three, probably because of the great time and energy requirements of child rearing as well as the enormous economic costs. Humans value freedom from economic constraints and freedom to enjoy the pleasurable aspects of life. In the past, efforts to secure these freedoms have almost inevitably been associated with producing a family with as many surviving children as possible. In the present, modern birth control technology permits us to satisfy our proximate drives without achieving the ultimate consequences of these drives, just as chocolate addicts exploit certain pleasure centres of the human brain that evolved in a much different context to serve different ultimate goals.

INTELLIGENCE AND PROBLEMS IN ITS MEASUREMENT

Before pursuing problems stemming from human behaviour, some discussion needed on the topic of intelligence. It would be impossible to arrive at a completely satisfactory definition of intelligence which takes into consideration all human populations in all their varied environmental setting. Most experts in the field would tend to agree, however, that intelligence does entail the ability to interpret symbols and to handle abstractions in a variety of ways. Many psychologists believe that the IQ test reflect this ability and hence intelligence. The IQ test was actually designed by Binet in France, early in this century, as one way to predict a student's success or failure in his studies in school. One reason the IQ test has become so well recognized is that performance on it does tend to indicate the degree of both academic success and occupational achievements later in life. IQ value is determined by obtaining the score which a person achieves on a test which supposedly denotes his ability to manipulate abstractions. The raw IQ test results for an age group in the population are then weighted in various ways (see below) to produce a so-called normal distribution or curve. Some persons are found at the extremes of low score and high score. Most, however, fall somewhere in between. The peak or average for IQ has been established as 100 points for the Caucasian population. The score for any one person is determined by relating the individual score to the average for the age group (Table 7.3).

Whether or not the IQ test actually measures intelligence is a matter of some debate, and space does not permit us to enter into details of the arguments. While IQ achievement serves somewhat to indicate accomplishments later in life, it must be noted that the very factors which determine academic achievement are correlated with those for occupational achievements

TABLE 7.3

Calculation of IQ. The means IQ for a given age group has been set as 100, since IQ's tend to fall into a normal distribution. The IQ for an individual is determined by dividing the mental age, as measured by performance on the IQ test, by the actual age and then multiplying by 100.

$$IQ = \frac{\text{Mental age}}{\text{Chronological age}}$$

Mental age	*Chronological age*	*IQ*
5	4	5/4×100=125
5	5	5/5×100=100
9	6	9/6×100=150
3	6	3/6×100= 50

in the society. It may not be intelligence itself which is being measured, but some ability to deal with factors upon which a particular society places great value. The question is just what is being measured related to the very construction of the tests. A disturbing fact is that the IQ scores from the population do not just happen to fal automatically into a bellshaped curve or normal distribution. Actually, the scores are forced to conform to this normal distribution by the very manner in which the examination is constructed.

Results have shown that boys perform better on test items dealing with mathematics, spatial relationships, and certain others, whereas girls surpass them in handling items dealing with language and memory. In order to prevent one sex continually scoring higher than the other on a test, let us say by the inclusion of more items dealing with mathematics which would favour boys, the items in the male-favouring category are balanced by a comparable number in the female-favouring

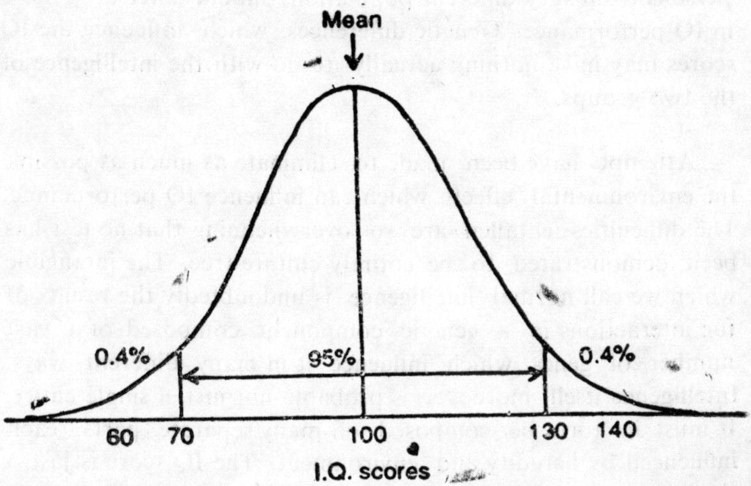

FIG. 7.5: When IQ scores are plotted for Caucasians who have been sampled, they tend to fall into a normal (bell-shaped) pattern of distribution. The mean IQ score has been set as 100 points for an age group. Approximately 95 per cent of the population have IQ's in the range of 70 to 130. Less than 1 per cent have IQ's lower than 60 and over 140.

group. The result of the selection of items eliminates high IQ performance which could be related to sex. A main point to note is that items can be selected if so desired to favour a certain group or to blur any intergroup differences which could be reflected in IQ test performance. We see that males and females of the same age group and from the same general environmental background can differ in IQ performance, depending on the items, but that this difference can be hidden by selection of the items in test construction. We have learned that different populations of humans differ in the frequency of certain genes and gene combinations. Therefore, two persons from two different population can be expected to have more genetic differences than two from the same population. Added to this would be very pronounced cultural differences. It would therefore seem that if two persons of different sex from the same population can differ on test performance, then two persons from very different populations should differ even more in IQ performance. Genetic differences which influence the IQ scores may have nothing actually to do with the intelligence of the two groups.

Attempts have been made to eliminate as much as possible the environmental effects which can influence IQ performance. The difficulties entailed are so overwhelming that no test has been demonstrated to be entirely culture-free. The intangible which we call normal intelligence is undoubtedly the result of the interactions in a genetic component composed of a vast number of genes which influence it in many different ways. Intelligence itself, moreover, is probably not just a single entity. It must in turn be composed of many separate parts, each influenced by heredity and environment. The IQ score is just a single numerical value which tell us nothing about the separate compartments, how they interact, or the relative importance of heredity and environment in each component. Therefore, while the IQ score does have some merit in predicting academic and later occupational success in a given society, it can still be asked: "What exactly is being measured, and what is the relative importance of hereditary and environment factors in IQ performance?"

The Use of Twin Studies in Genetic Analysis

Whatever the full significance of IQ performance, there is no doubt that it does entail a hereditary component. The estimate that portion of IQ which rests solely on genetic factors, investigations have relied heavily on studies of identical twins. Identical twins (monozygotic or MZ) have all of their genes in common; their genotypes are identical. Fraternal twins (dizygotic or DZ), like any other brothers and sisters, have no the average 50 per cent of their genes in common. Since the genotypes of identical twins are the same, monozygotic twins provide a superb opportunity to help evaluate the relative roles of environment and heredity in the expression of a trait. A truly unique opportunity is provided by those rare identical twin who have been raised apart since infancy or early childhood. Since the environmental effects would be more varied on twins reared apart than on those raised together, the contribution of environmental factors to the expression of a trait can be revealed. A pair of identical twins thus gives the opportunity to study the expression of the same genotype in a similar environment (monozygotic twins reared together) and the expression and the same genotype in different environments (MZ reared apart).

Studies of dizygotic twins (DZ) and ordinary brothers and sisters reared togetner and reared apart are also related to the data from identical twin studies. Two identical twins should always show the same trait if that trait is based entirely on hereditary factors. Such a trait would be the blood type. Monozygotic twins always agree in their ABO, Rh, MN, or other blood groupings. We say that 2 members of a twin pair are concordant if they are alike with regard to a certain trait. They are thus concordant if they both exhibit a trait (such as the same blood type) or if they are both free of a trait (neither is an albino). Among identical twins, 100 per cent would be concordant with regard to ABO blood type. Members of a twin pair are said to be discordant if they differ with respect to a given trait. If 1 pair of identical twins in a study is discordant, the operation of environmental factors in the expression of the trait under study is indicated. If MZ and DZ twins are very similar in concordance value for a trait, then the strong role of the environment is indicated, since the members of each dizygotic twin pair have only about 50 per cent of their genes in common. If based largely on a hereditary component, the MZ twin pairs should show a much higher concordance value. We can thus use the extent to which twins, identical and fraternal, differ in their concordance values as a sort of measure of the effects of the environmental and genetic components on the expression of a characteristic or trait. Suppose 50 pairs of identical twins are studied for a trait and that 40 of the pairs (80%) show concordance. A study of 50 fraternal twins for the same trait shows that 10 pairs (20%) are concordant. This would provide evidence for a strong hereditary component. If

		Identical Twins						*Fracternal Twins*				
		20	*40*	*60*	*80*	*100*		*20*	*40*	*60*	*80*	*100*
Beginning of sitting up	(63)		82%				(59)		76%			
Beginning of walking	(136)		68%				(128)	31%				
Hair colour	(215)			89%			(156)	22%				
Eye colour	(256)			99.6%			(194)	28%				
Blood pressure	(62)		63%				(80)	36%				
Pulse rate	(84)		56%				(67)	34%				
Handedness (left or right)	(343)			79%			(319)		77%			
Measles	(189)				95%		(146)		87%			
Clubfoot	(40)	32%					(134)	3%				
Diabetes mellitus	(63)			84%			(70)	37%				
Tuberculosis	(190)			74%			(427)	28%				
Epilepsy (idiopathic)	(61)			72%			(197)		15%			
Paralytic polio	(14)	36%					(33)	6%				
Scarlet fever	(31)		64%				(30)		47%			
Rickets	(60)			88%			(74)	22%				
Stomach cancer	(11)	27%					(24)	4%				

Mammary cancer	(18)	6%	(37)	3%
Cancer of uterus	(16)	6%	(21)	0
Feeblemindedness	(217)	94%	(260)	47%
Schizophrenia	(395)	80%	(989)	13%
Manic-depressive psychosis	(62)	77%	(165)	19%
Mongolism	(18)	89%	(60)	7%
Criminality	(143)	68%	(142)	28%
Smoking habit	(34)	91%	(43)	65%
Alcohol drinking	(34)	100%	(43)	86%
Coffee drinking	(34)	94%	(43)	79%

FIG. 7.6: Concordance and discordance. The shaded areas indicate the percentage of concordance among monoxygotic (MZ) and dizygotic (DZ) twins. When both members express a certain trait or when both members of a twin pair are very similar with regard to a feature which does not have distinct traits, they are considered concordant. The percentage of discordance can be estimated by the unshaded areas and is equal to 100 minus the percentage of concordance. The numbers in the parentheses indicate the number of twin pairs studied for each item.

both groups, monozygotic and dizygotic, showed approximately the same amount of concordance for a trait (or conversely, the same amount of disordance) such as 70 per cent, then the greater importance of the environment would be indicated.

A list of conditions along with the per cent of concordance. Note from the table that eye colour shows a much higher concordance value between MZ than between DZ twins. This tells us that the genetic similarity of the MZ twins is highly significant in the expression of this condition and that eye colour has a strong hereditary component. The Fraternal twins show much lower concordance, since members of a pair will not always have the same genes for a trait because only about 50 per cent of all their genes will be common to both. Note that susceptibility to measles is not so different when concordance is compared between the 2 classes of twins. This is a reflection of the greater magnitude of environmental influence on this condition. The values for some conditions, such as stomach cancer, suggest a hereditary component significant difference between MZ and DZ twins in concordance value) but a large environmental component (73% of the identical twins are discordant).

Heritability and the Intelligence Controversy

There are certain limitation in twin studies which must be recognized. For example, the operation and impact of prenatal environmental influences on both MZ and DZ twins is hard to evaluate, and these could operate to influence the expression of a condition even when twins are separated from birth. If could thus be a prenatal environmental factor which causes a concordance; and this would be measured as a genetic one. Nevertheless, valuable information can be gathered from twin studies on the relative roles of environment and heredity in the expression of a trait or characteristic. The information can be used to measure heritability. Heritability is a measure of that percentage of the variation shown in the expression of a condition or characteristic which is due to genetic factors. Any complex characteristic such as height or weight which involves many genes and environmental factors will exhibit variation. If all that variation could be shown to be due entirely to genetic factors, then the heritability would be equal to 1. This would mean that the variation seen from one individual to the next is due completely to genetic influence. If the heritability were 0.5, then half of the variation seed is due to genetic factors the other half to environment effects. A heritability of 0 would tell us that all the variation is due to the environment.

Intelligence is a most complex characteristic, and it is well known to vary greatly from one person to the next. While we do not know exactly what the IQ performance measures, it would appear that it does measure some kind of mental ability. IQ has been used as an index of intelligence in numerous studies, and the results leave no doubt that intelligence, as measured by IQ performance, has a significant hereditary component. IQ performance of identical twins, some reared apart, other together, is compared to that of fraternal twins and to that of related and unrelated persons. If heredity alone were involved, the correlation between monozygotic twins should be 1.0 and between fraternal twins and ordinary siblings, 0.5. The correlation between MZ's is significantly higher than that for the other groups. The difference between MZ's together as opposed to MZ's reared apared indicates an environmental influence. From analyses using this approach, some investigators have estimated that environment accounts for approximately 20 per cent of the variation seen in a population in IQ performance. In other words, the heritability of intelligence as indicated by IQ is considered by some students of the subject to be in the vicinity of 80 per cent. This would mean that 80 per cent of the variation shown in IQ is the direct result of hereditary factors. The procedures used to arrive at a heritability value of 0.8 are somewhat complex and entail a number of assumptions. While there is certainly a significant genetic component involved in the variability seen in IQ performance in a population, it seems highly questionable to assign a rigid heritability value of 0.8 or 80 per cent.

Relevant to this point is the concern which has been shown in some quarters concerning the lowering of intelligence due to unequal reproductive rates among the various economic groups in society. Some have feared that the higher rate of reproduction of persons in lower economic strata, who tend to score lower on IQ than those from higher socio-economic classes, will lead to eventual decrease of intelligence in the population. There have been few tests designed to test different generations in order to determine whether or not intelligence does vary from one generation to the next. Results from the few tests that have been conducted indicate a significant increase in test scores! Besides arguing against fears of lowering intelligence, the results seem to show a very rapid change in test score performance in just one generation or less. This can hardly be due to a genetic change of monumental proportions. Changes in environmental factors are strongly indicated, and these would appear to be instrumental in influencing the test scores and causing change in a short period of time.

Observations such as these lead us to be cautious when assigning a rigid value to heritability, a value when often rests on many assumptions. Heritability values for many characteristics have been shown to vary from one population to another similar population in a different environment. Heritability data on IQ have been obtained largely from studies of white populations in various regions of the United States and Europe. Since heritability for any characteristic is a figue derived from a population in a particular environment at one particular time, we may question the application of a fixed value for heritability of intelligence to one population when it has been calculated from data obtained from studies on another. Yet this is what appears to be the case in studies where conclusions have been drawn about racial differences in intelligence based on IQ test performance.

Few would deny that scores achieved by American blacks on IQ tests tend to be lower than those for American whites. Several studies suggest that the average IQ score for the white population is 100, where that for the black is 85. Even if this difference between mean IQ values should prove to be the case, it must be understood that many blacks score higher than the average for whites and many whites score lower in test performance than the average black person. No one can say that any black must score lower than a white person. While a difference between an average score of 100 and one of 85 is mathematically significant, the interpretation of this difference has led to many heated debates. According to many, this difference reflects various disadvantages experienced by American blacks in educational and economic conditions. Little is known about the subtle effects which nutritional deficiencies, even those before birth, can have on a character as complex as intelligence. Perhaps all of the difference in IQ performance between the two groups stems from environment alone. Moreover, this performance may not be related to intelligence at all.

There are others who hold an opposing view, citing a heritability of 0.8 to which they rigidly adhere. These persons argue that 80 per cent of the difference or variation seen in IQ achievement has a hereditary basis. Only 20 per cent at the most is contributed by the environment. Jensen and other proponents of this viewpoint argue, therefore, that only 20 per cent of this variation could eliminated if all environmental differences were eliminated. Under exactly the same environment, there would still be a great variation between the black and white groups, and this would be entirely genetic. Assuming a heritability of 80 per cent it can be shown mathematically that elimination of the variation due to environment would raise the average IQ score for the black population only 1.6 points. The end result would be only a slight increase in IQ performance for American blacks. Therefore, the argument goes, the mean IQ value of the whites is truly higher than that of blacks and is largely due to a difference in hereditary factors which affect inteligence.

This sort of reasoning has been used by many well-intentioned investigators who are striving to improve educational and occupational achievements for all people. It has, of course, also been seized upon eagerly by those who wish to prove that one race is more intelligent or better than another for their own unscientific reason. While the IQ controversy has led to many unfortunate debates and ill-founded statements, little can be accomplished by those persons of any persuasion who choose to avoid discussing intelliegnce entirely. Value for everyone can be derived from gaining insights into normal intelligence and the environmental and genetic factors which do moid it. Any differences between groups on some test such as IQ should not be construed to mean that any person or group is better in some way than any other. A test should be treated as one device which may enable the investigator to obtain a clue on some aspect of natural ability. It should not be taken to mean that it must reflect a person's innate intelligence, a chracteristic composed of so many components. The subtle effects of environment must never be overlooked in the evaluation of such a complex attribute. Strict adherence to some mathematically computed value for a components of a complex character should be avoided until it can demonstrated that such a value is constant under all conditions. A figure such as heritability can be considered truly reliable only when obtained from studies conducted on similar and very different populations under the widest assortment of environments. This has not been done. There is thus a complete lack of data on heritability differences seen within a group as compared to heritability differences which might pertain to separate groups of populations under a range of environmental backgrounds.

No true scientific investigation even sets out to prove a point, nor does it use vague or ill-defined mathematical concepts to supports a chosen argument. The most effective person is the one who gathers known facts, evaluates them, and tries to apply them constructively. He or she may use them to argue successfully against those who are ill-informed, unaware of pitfalls in their reasoning, and especially those blinded by emotion.

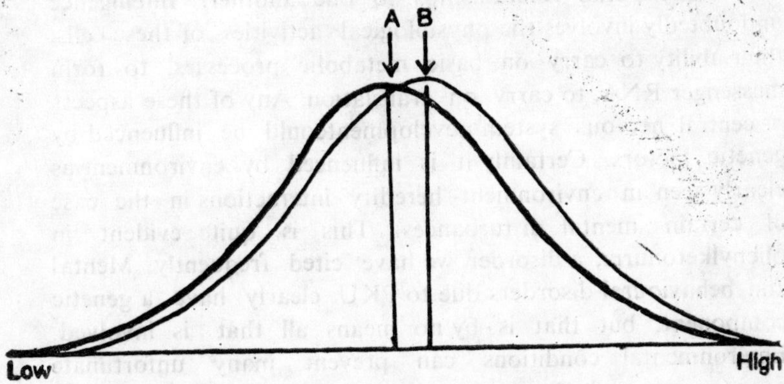

FIG. 7.3: Theoretical mean IQ differences between two races. Assume that two races, A and B, are truly found to differ significantly in mean performance on IQ tests. In such a case, there would be a very large range of overlapping between the two races. Many persons of race A, the one with the lower mean IQ, would nevertheless score much higher than the average IQ of race B, even though the latter race has the higher mean score. The differences found within any one race are greater than those between the two races.

Genes, Environment and Mental Capacity

Another point to bear in mind concerning any measure of intelligence is that any test score is a single numerical figure. As we have noted, intelligence is not just one entity. Rather, it is a complex of interrelated parts, some of which may have a high degree of independence. When we say that intelligence has a polygenic basis, meaning that many genes can influence intelligence, we do not mean that many genes contribute certain units of something called intelligence. It would be absured to think of intelligence genes, each one adding a certain degree of intelligence in the way that genes which control melanin production influence pigmentation. Instead, there are unknown numbers of genes which can influence development of the brain and other parts of the central nervous system in an unknown variety of ways. Normal intelligence must depend on a certain number of nerve cells and their spatial relationships to one another. Intelligence undoubtedly involves the physiological activities of these cells; their ability to carry on basic metabolic processes, to form messenger RNA, to carry on translation. Any of these aspects of central nervous system developmentcould be influenced by genetic factors. Certainly it is influenced by environment as clearly seen in environment- heredity interactions in the case of certain mental disturbances. This is quite evident in phenylketonuria, a disorder we have cited frequently. Mental and behavioural disorders due to PKU clearly have a genetic component, but that is by no means all that is involved. Environmental conditions can prevent many unfortunate consequences of the untreated disorder. The recessive PKU gene is thus one which can have an effect on intelligence, but we now have some understandin g of what this means from the knowledge which has accumulated on the molecular basis of the syndrome. Certainly, the normal allele of the PKU recessive is important in the development of normal intelligence, but we see that its effect stems from the control of a specific biochemical step. It would be ridiculous to say that it contributes a unit of intellignce. Many normal genes are required for expression of normal intelligence, since a defective allele many cause a derangement in cellular chemistry which in turn may lead mental retardation. An unfortunate victim of Huntington's disease may be a person of high intelligence who deteriorates mentally as a result of central nervous system deterioration later in life. The allele of the defective dominant gene is thus required for the maintenance of normal mental capacities, but again we see clearly that no unit of normal intelligence or unit of behaviour is involved. Instead, human intelligence and human behaviour are divided into many components. A gene or an envirnomental disturbance of some kind may influence one or more of them and operate against what is considered the normal expression of these human characteristics.

A lengthy list could be compiled of single-gene defects which are known to influence mental development and which can cause behavioural disturbances. Not one of these produces more bizarre effects than the sex-linked recessive associated with Lesch-Nyhan syndrome, a kind of cerebral palsy. An early symptom of this disder is vomiting. However, this may be associated with the aggressive behaviour so typical of the syndrome. Performance on intelligence tests indicates that victims of this disorder are retarded, but many feel that these children possess a good level of intelligence. They are certainly bright in comparison to most persons who are institutionalized for retardation. Children with the disease develop a compulsion to mutilate themselves, bitting the lips, tongue and fingers. When upset, they may vomit, covering those about them. Speech is greatly impaired in the Lesch-Nyhan syndrome, and spastic movements are evident. Particularly poignant is the observation that affected children apparently desire to be protected against the mutliation they tend to inflict on themselves. They allow the fitting of restraints (mittens, splints) to protect them. The child tends to become relaxed with the restraints; when they are removed, he cries out, seeming to fear his own tendency to mutilate himself. There is some reason to believe that stress may trigger this compulsion. Besides harming himself, the Lesch-Nyhan child is extremely aggressive. He may kick, hit, and pinch persons caring for him. Still, he may apologize for his aggressive behaviour!

There is variation in the expression of certain of the perverse behaviour traits. Onset of self-mutilation was delayed until the age of 14 in 1 case, and 2 older children did not express this tendency. The variation may be a reflection of the amount of biochemical upset in a particular case. The molecular defect in the Lesch-Nyhan syndrome is known, and information on this very rare sex-linked disorder has great relevance to the study of aggressive and compulsive behaviour in general. It raises the question of the role of genetic control in the expression of excessive aggression and in the manifestation of several types of compulsiveness. The recessive gene has been identified with a deficiency of a certain enzyme (hypoxanthine-guanine-phosphoribosyltransferase or HGPRT). This enzyme is involved in the metabolism of purines and is either absent or occurs in an unstable or inactive form in those children with the mutant gene. Lack of the normal enzyme leads to an accumulation of a substance which, when present in excess, causes an increase in the metabolic rate. How the enzyme defect is related to the aberrant behaviour is unknown, but it apparently causes biochemical derangements which in turn are responsible for neurological damage. The biochemical upset in the Lesch-Nyhan syndrome also occurs in patients who suffer from gout, even though they exhibit no neurological damage. Like gout sufferers, Lesch-Nyhan children secrete excessive amounts of uric acid in the urine. This symptom may be apparent in a 6-9 month old child before some of the neurological disturbances become evident. Lesch-Nyhan victims eventually develop gout symptoms, and they also tend to suffer kidney damage, including the development of stones. Kidney damage is usually the responsible cause of death before reproductive age. Since the trait is sex-linked, no male can transmit the gene. Consequently, a female victim cannot arise, since such an individual must receive a defective gene from the paternal as well as the maternal parent. Attention undoubtedly will continue to be paid to this rare disease for the information it can provide on problems of arthritis, and especially for the light it can shed on behavioural disturbances and the biochemical upsets which may trigger it.

Other single genes are known (autosomal, sex-linked, dominant, and recessive) which exert their primary effects in very different ways but which can bring about mental retardation as well as certain behavioural disturbances.

Heredity and Environment in Mental Illness

Mental illness is a matter of great concern, since it takes such a toll in the burden it imposes on large numbers of the population. The psychoses are the most serious of the mental illness, and the most frequent psychosis is recognized as schizophrenia, characterized by withdrawal into the victim's personal world which is pervaded by very bizarre thoughts and gives rise to most unusual and most unexpected behavioural responses. Schizophrenia exists in several forms, making recognition and accurate diagnosis difficult. This has presented a problem in the elucidation of the genetic component associated with this psychosis. Some psychiatrists have held the view that the disorder stems solely from environmental causes. However, present data show that a significant genetic component exists. Concordance values for identical twins are much higher than those for fraternal twins or for unrelated persons. Identical twins do show discordance; therefore, the environmental influence must be recognized as a significant factor in the expression of the psychosis. However, the importance of the genetic component is seen not only in the higher concordance between monozygotic than between dizygotic twins, but also from other kinds of observations. For example, children of schizophrenic parents have been separated from the mentally ill parents at birth and raised in normal families. The risk of developing schizophrenia, however, in such children is only slightly less than if they had not been separated. Comparison with other adopted children has shown that the act of adoption itself is not the causative stress which brings the schizophrenia to expression.

There has been much debate concerning the responsible genetic factors which can predispose toward schizophrenia. Some investigators favour a single-locus hypothesis in which a dominant gene is implicated; others favour a recessive-gene interpretation. According to some adherents of the latter, the predisposing recessive would produce an added risk of schizophrenia even in the heterozygote. Most schizophrenics would be homozygous for the recessive gene, but according to this hypothesis, unknown environmental factors can exert a strong influence on expression of the genotype. Carriers of the gene in both the homozygous and heterozygous conditions would thus run a greater risk of developing schizophrenia than would a person lacking the gene entirely. However, persons with the gene, including the homozygote, could be normal. On the other hand, even those lacking the predisposing gene could develop schizophrenia solely as a result of certain environmental stresses.

There are other students of the problem who support the concept of a polygenic rather than a single-locus hypothesis. Environmental influence is also recognized by adherents of this hypothesis. Whatever, the exact nature of many defective genetic factors in schizophrenia, no molecular or biochemical disturbances have been identified, although some derangement in biochemistry which affects nerve hormones (neurohumors) is anticipated. The kind of environmental stress which can trigger the psychosis is also unknown.

Most of us have heard of the manic-depressive psychosis, another fairly common mental illness characterized by extreme changes in mood which sway from deep depression to great elation. A genetic component is also indicated for this condition, and it is a different one from that responsible for schizophrenia. Again there is a debate on the nature of the genetic component which is involved.

Studies of the genetic and environmental components in mental aberrations are very important, not only for the information they may yield to lead to successful treatment and prevention, but also for the information which may be provide in an understanding of other, more subtle aspects of behaviour. Twin studies indicate that a genetic component may exist for personality type. The data show that personality can be greatly influenced by the environment, much more so than for certain other complex characteristics such as height, weight, or IQ achievement. But the possibility that heredity may be involved in the determination of personality type is also suggested. This has a great bearing on the study of criminal behaviour, a very controversial topic which becomes very heated when discussed in relation to any hereditary predisposition. No definitive statements can be made at the moment, but the possibility cannot be dismissed that some genetic factors exists as well as environmental ones which predispose to criminal behaviour.

Ethical Considerations in Genetic Investigations

We have already discussed the possibility of an extra Y chromosome as a predisposing factor toward criminal or aggressive behaviour. This topic leads us to another controversial matter, that of the ethics and morals in genetic investigations and procedures of a certain nature. Proposed studies of children with an extra Y chromosome have focused attention on the type of heated debate which can surround genetic research. More and more, we can expect to encounter problems which pose questions with serious ethical ramifications. The relationship of an extra Y chromosome to antisocial behaviour is still unclear, and much more clarification is needed before positive statements can be made. The discovery at Boston Hospital that the XYY conditions occurs with a higher frequency than formerly suspected (perhaps 1 male birth in 1000) indicated to some a rather urgent need to accumulate more information on the condition in order to settle the matter. A study was undertaken to follow up children with XYY and those with normal XY chromosome constitutions and to observe the two groups throughout a period of years. Any behaviour differences between them would be noted. Such followings, it has been argued, could yield several benefits. Any behavioural aberrations might be detected at an incipient stage when they could yield more effectively to treatment and guidance moreover, observations of those XYY children who happen to manifest no behaviour problems could be related to studies of aggressive XY as well as to aggressive XYY children. The information obtained could cast light on those environmental influence which may mold a child's behaviour or personality. Personnel the extra Y may prove to contribute little or no effect.

The study has been challenged by at least one group of scientists who maintain that it is unethical and unscientific. Most important, it is argued, such a study is potentially harmful to the children involved. It immediately sets any child in the study apart by calling attention to him. Moreover, parents with an XYY child, knowing he has some kind of genetic anomaly, will inadvertantly treat him in a different fashion, which could distort has entire mental development. This in turn could set off a serious psychological disturbance which could or could not be related to the presence of the extra Y. Many who argue against the study feel that it focuses attention on very unusual genetic conditions and thus detracts from the socio-economic problems which are known to contribute to antisocial behaviour and criminality.

Many proponents of the investigation admit that it may be better not to allow the XYZ person to know about his genetic constitution, since he might conceivably develop normally without this knowledge. If the design of the study could be altered to prevent this, it then raises the provocative question of a person's right to know and the right of parents to have such information if it is known to medical personnel. The controversy has been sufficiently heated to cause a cessation in further identification of XYY males. However, the study will continue on groups which are presently under investigation.

The XYY story illustrates the kind of moral dilemma which will continue to arise in genetic research as the procedures used in detection and treatment of genetic disturbances become more sophisticated. No one would argue that steps should be avoided which could prevent human suffering. The dilemma which arises often centres around the problem of whether the procedure will cause in the long run more problems for those it tries to treat, as well as greater problems for the human population in general.

Another type of research which is becoming more publicized involves research on the human fetus. Many have attacked such measures on ethical grounds, while others applaud it for the advances it can achieve in medicine. A study made for a national commission which will advise the department of Health, Education, and Welfare on fetal research has reported that total abstention from this kind of study would have caused delays in our advance of medical knowledge and would have cost thousands of lives. One area which profited, so the study reports, is that of the treatment of babies suffering from blood disorders due to Rh incompatibilities. Many thousands of infant deaths would have resulted with no research on fetal material, since animal research would not have been as satisfactory. Moreover, without the information obtained, many additional thousands would have become seriously incapacitated due to brain damage from Rh incompatibilities, placing burdens as well on society. As many as 12 per cent of married couples may face some kind of risk due to Rh incompabilities. The development of a vaccine which prevents the Rh negative mother from becoming sensitized by Rh+antigens has been accelerated greatly by research utilizing fetuses.

The report also pointed out that a complete ban on research involving fetuses would have prevented the development of amniocentesis, a procedure in almost routine use to diagnose fetal disorders as well as to detect genetic defects in utero. In some cases, the parents are presented with the option of abortion to prevent the birth of babies doomed to years of suffering.

Genetic Engineering and Problem Society

The problems which will continue to confront society in areas of genetic research underscore the need to weigh carefully the benefits and the dangers entailed in a particular approach. The best course to take may, unfortunately be unknown at the moment a decision is required. Urgency for some guidelines in certain types of genetic research is perhaps best illustrated by the decision reached at a conference of biologists assembled to debate research in genetic engineering, an area which involves the manipulation of genetic material by the insertion or elimination of DNA into a cell. An ideal solution to a genetic disorder would be the replacement of a defective gene by its normal allele in those cells in which activity of the normal gene is required. Accomplishment of such a feat is not in the immediate future and depends on research in various directions, among them the assignment of human genes to precise locations on the DNA of the chromosomes. Complete cure would require the insertion of the normal allele into germ tissue so that correct information would be transmitted to the next generation. While still a distance from effecting such complete cures, certain approaches have already been undertaken. It has been well established that some bacterial viruses can incorporate into their DNA bits of genetic information from a host cell and then transfer these to another bacterial cell which the virus later infects. Actual attempts have already been made to infect humans with viruses carrying genetic information missing from the humans' cells. Success has not been achieved, although enzyme-deficient human cells in vitro have been shown to acquire the ability to manufacture a specific enzyme after infection with viruses carrying the required functio al gene. It is still unknown whether the viral genetic material with the added gene has been incorporated into the chromosome or whether it is residing elsewhere in the cell. Another question which arises is whether infection of a person with a virus would in the long run be detrimental. Since the DNA of the virus may persist in the human cell and could very possibly undergo transcription leading to disturbances in the physiology of the cell. A safer approach might be the injection of the enzyme known to be lacking in a disorder, as has already been reported with some success in the case of Gaucher's disease. But this approach does not correct the genetic information and could therefore require continued treatments.

The isolation of specific genes has actually been accomplished in the laboratory, as has the storage of particular genes from certain lower animals. Isolated genes can provide the geneticist with the opportunity to decipher the actual makeup of a gene and to study the mechanism of gene action and control. However, it is just this kind of approach to genetic engineering which has led to the conference of concerned biologists. Techniques have now been perfected which enable the transplantation of genes from one kind of animal into the DNA of another. This can permit the storage in a simple cell type of genes derived from a complex cell. This feat has been made possible by the use of some enzymes which were found to be capable of breaking DNA into segments and of others which can repair the breaks. Segments of DNA have been taken from animal cells such as those of the sea urchin or the South African toad. The segments have then been inserted into bacterial DNA through the action of enzymes which can recognize and break specific regions of the DNA. The foreign animal DNA can then be inserted into DNA of the simples cell, and all breaks can be healed. Such bacteria then reproduce and form a storehouse of specific animal genes for study as needed. Large amounts of products of the inserted genes could pile up in the altered cells. Another advantage would then be the production in the simpler cells of large amounts of a valuable substance (insulin perhaps) required by persons with serious genetic diseases.

The potential dangers of this kind of genetic surgery for the human have caused alarm within scientific circles. It is conceivable that transfer of foreign genes from higher cell types of common bacteria such as E. coli could impart very infectious properties to the microorganism which could then run rampant, causing novel kinds of infections. There is even the danger of tranfer to the human of known cancer-inducing animal viruses from cells in which virus genetic material may be stored. The potential for harm in this field of research has prompted the scientific world to recommend guidelines in the pursuit of such studies. The guidelines have no legal basis, but the moral compunction to follow them should be strong, since so many illustrations scientists have recognized the need for them. The proposal of safeguards and guidelines is very unusal in the scientific world, which has tended to avoid the imposition of resiraints on investigations along certain pathways of inquiry. We see in this case complete intellectual freedom being weighed against the dangers inherent in a particular approach. To many, restraints appear preferable during the pursuit of an investigation whose capacity for overwhelming harm is at the movement unknown.

Our understanding of human genetic disorders will continue to present on end of ethical dilemmas. There are many who seriously question the morals of any kind of abortion, even when the fetus is known to be defective and the victim of an unfortunate disorder, such as Tay-Sachs disease or Down's syndrome. Perhaps even more disconcerting are problems which relate to the successful treatment of genetic disease. The life of an affected person may be spread, but the treatment required may be lifelong and experience. It may imparted an excessive burden to the victim, the family, and the medical teams which are required. There are those who ask what benefits are gained by saving from early death those children with genetic defects which require expensive treatment that permits survival for just a few additional years. Could the efforts be better utilized in another direction even though death in certain cases would be a certainty if the disorders are left untreated? Remember that many genetic disorders duo to single-gene defects are very rare. A more and more continue to be understood and treated, will this divert attention to an ever-growing list of disorders, each affecting few people? Will this shift attention, thus restricting the resources and efforts needed to solve health problems which affect very large numbers of people?

Not to be overlooked is the problem presented by saving persons with genetic disorders who will later reproduce and pass on the defective genes. Phenylketonuria illustrates more than one problem of this kind. No one would question the value of treatment of PKU victims. However, when a PKU woman saved from mental deterioration, becomes pregnant, additional steps must be taken to protect the child she is carrying. This is so, since the PKU adult will very likely be on a normal diet. Procedures must be undertaken to control the diet of PKU mothers in order to avoid fetal damage as a result of high levels of phenylalanine in the maternal blood.

There are those who point to possible dangers to the population as more and more persons with defective genes reproduce and pass the defective genes into the population gone pool. Many geneticists consider it possible that a crisis will eventually arise in future generations if freedom to reproduce is left unchecked. Others answer this by saying that while increase of a dominant gene would be quite rapid (about 6 times the original frequency in 5 generations), the frequency of the more common defective recessives would rise very slowly (only about 1 % in a generations). Man's knowledge in the long run, so this argument goes, would enable him to devise ways throughout the years to cope with the increase in the frequency of affected persons. In the future, persons who are now considered defective in some way would impose no burden to the population and incur no disadvantage and result of the perfection of procedures and treatment which are at the moment unimaginable.

In the long run, it becomes a question of human behaviour, that multisided characteristic composed of so many intangibles The human's individual behaviour can certainly result in harm to himself. Should restrictions be imposed upon individual behaviour if it affects only the individual who practics it? The use of certain drugs is known to be harmful to the person. Even certain drugs known to be benefical to the average person in the treatment of illness may cause violent reactions in some indivduals, as seen clearly in the case of G6PD- deficient persons to primaquanie and other substances. Such people are normal in all respects other than their drug response. Another example is seen in the violent reaction to barbiturates experienced by those with one or another form of porphyria, a group of disorders with different genetic bases. Some of these people are essentially normal but may exhibit varying degrees of sensitivity to hight. Exposure to barbiturates, however, can result in death. Many more examples of drug anomalies could be given. It should at least be appreciated that normal persons may respond differently to a given drug, in their obvious physical reactions to it and even in their ability to respond in any way at all. More and more drug anomalies are certain to appear as more therepecutic agents are designed.

Since harmful drug responses with a genetic basis are well known, what should be practiced to regulate personal use of substances whose effects are poorly defined? Should a marijuana user be punished because some evidence indicates that the immune response is lowered with habitual marijuana use? Perhaps just the dangers of usage should be publicized, as in the case of cigarette smoking, and personal choice permitted. But what about evidence that marijuana may depress the level of the male hormone in habitual male users of this substance? In mice, rats, and even in some primates there is experimental evidence which indicates that interference with male hormones during fetal development can interfere with the completion of normal male development, including behaviour. While still an unknown area, the question arises whether marijuana use by a pregnant women could adversely affect the development of the male fetus. Other reports suggest that marijuana use interferes with the assembly of RNA and even of amino acids into proteins. Such possibilities raise the point that personal use of substance may do more than harm the individual user alone. The potential for harm to future generations by any mutagenic action of drugs in current use is also unknown but cannot be disregarded.

It must be kept in mind that the personal behaviour of each human can affect the welfare of others. The behaviour of the human species collectively can alter the existence of other species on the planet. Research into the genetic and environmental components of the human's most intricate characteristic, behaviour. may hold significant bearing on the future of Homo sapiens and all other forms of life. Whether the human's genetic potential can be put to its best use for the survival and advancement of all life forms will finally yield an answer to the tantalizing question, "Is today's human truly a higher form of life?"

HUMAN BEHAVIOUR AND EVOLUTIONARY THEORY

We have at the threshold of an exciting new phase of the analysis of human behaviour, and a new generation of biologits and social scientists has an unparalleled opportunity to help develop this phase. Evolutionary theory has just begun to be used to explain the diversity of human behaviour among cultures. As we have seen, gene thinking generates a host of predictions about how humans should behave under variable ecological conditions. The extraordinary differences in the behaviour of humans of different cultures can be used to test directly whether there is in fact a correlation among cultural strategy, the ecology of the society, and the inclusive fitness of members of the culture. Weighed against human cultural diversity, the variation in the social behaviour of other species is vanishingly small. It is unfortunate that much of this diversity is disappearing before it can be thoroughly described and before new information can be collected that would permit scientists to test evolutionary predictions about human behaviour. Nevertheless, there is still time for a productive union of anthropology, psychology political science, economics, sociology, and evolutionary biology although forming such a union will be a challenging task. We have a great chance of increase our knowledge of ourselves; I hope that have the willingness to do so.

8

The Ecology of Desert Plants

What selects the plants in an extreme environment? Is it a sort of war? The study of plants in the desert and in the laboratory indicates otherwise, which points a moral for the human species.

The laws of human behaviour are very much in dispute. largely because there are no obvious experimental approaches to them. But animal and plant behaviour can be studied both in nature and in the laboratory, and the science of their ecology should eventually be helpful in the understanding of human relationships, for the basic laws which govern the interrelations among organisms in general must also underlie human behaviour. Ecology is an extremely complex study. For a relatively uncomplicated case, from which we may be able to extract some generalizations about behaviour, I want to take you with me to the desert to survey its plant life.

The desert is an ideal area for research. It is usually unspoiled by the encroachment of civilization. Its plant life is sparse enough to be studied conveniently in detail, and it shows clearly and primitively the effects of the physical factors at play in the environment. Most important of all, the desert climate is violent: winds sweep over it unchecked, and its temperature and rainfall swing between wide extremes. Rainfall may vary fivefold from year to year. There are so few rainstorms that the effects of individual rains can be measured. The desert's sharply contrasting conditions can be reproduced in the laboratory for convenient experimental investigation of the germination and growth of plants. And the desert has an unending lure for the botanist; in the spring it is a delightful place.

The most extreme desert in the U.S. to Death Valley. Screened off from the nearest source of water vapour—the Pacific Ocean—by the tall Sierra Nevada, the valley bottom has an average annual rainfall of only 1.35 inches. It has almost no surface water—only a few springs bringing up the scanty runoff from the dry surrounding mountains. Since it is sunk below sea level, Death Valley has no drainage. As a basin which holds and collects all the material that may be washed into it from the mountain canyons, it has accumulated salts in its central part. Seen from above, this salt bed glistens like a lake, but a traveller on foot finds it a dry, rough surface, studded by sharp salt pinnacles which crackle and tinkle as they expand or contract in the heat of the day and the cold of the night.

In the salt plain no green plants can grow; there are only bare rocks, gravel and salt. But on the fringes of the plain plant life begins. Here and there are patches of a lush green shrub—the mesquite. With their tender green leaflets, which suggest plenty of water, the plants seem completely out of place. Actually they do have a considerable source of water, but it is well underground. The mesquite has roots from 30 to 100 feet long, with which it is able to reach and tap underground lenses of fresh water fed by rain percolating down from the mountains.

The mesquite is the only shrub that can reach the water table here with its roots. But a mesquite seedling must send its roots down 30 feet or more through dry sand before it reaches this water. How, then, does it get established? This is one of the unsolved mysteries of the desert. Most of the mesquite shrubs in Death Valley are probably hundreds of years old. Some are all but burried by dunes of sand, piled around them over the years by the winds that sometimes blow with great force through the valley. There are places where dozens or hundreds of stems protrude from a dune, all probably the offshoots of a single ancient shrub rooted beneath the dune.

Another Death Valley plant endowed with a remarkable root system is the evergreen creosote bush. It has wide-reaching roots which can extract water from a large volume of soil. The creosote bush is spread with amazingly even spacing over the desert; this is especially obvious from an airplane. The spacing apparently is due to the fact that the roots of the bush excrete toxic substances which kill any seedlings that start near it. The distance of spacing is correlated with rainfall: the less rainfall, the wider the spacing. This probably means that rain leaches the poisons from the soil so that they do not contaminate as wide an area. We commonly find young creosote bushes alone roads in the desert, where the road builders have torn up the old bushes.

During prolonged periods of drought creosote bushes lose their olive-green leaves and retain only small brownish-green leaves. Eventually these also may drop off, and the bush then dies unless rain comes soon afterward. However, it takes a really long drought to kill off all the creosote bushes in an area. They have suffered severely in some areas of the southern California deserts during the drought of the past five years. Because a killing drought tends to remove them wholesale, there are usually only a few age classes of creosote bushes in an area; each group springs up after a drought or during a period of unusual rainfall.

There are other shrubs that master the harsh conditions of the desert, among them the lush green Peucephyllum, which seems to be able to live without water, and the white-leaved desert holly, which grows in fairly salty soil.

Two prime factors control the abundance and distribution of plants: the number of seeds that germinate, and the growing conditions the seedlings encounter while they seek to establish themselves. In the case of the desert shrubs the main controlling factor is the growing conditions rather than germination, for though many seedlings may come forth in a rainy season, few survive long enough to become established. The story is entirely different for the annual plants in the desert.

There are years when the desert floor in Death Valley blooms with a magic carpet of colour. In the spring of 1939 and again in 1947 the nonsalty portion of the valley was covered with millions of fragrant, golden-yellow desert sunflowers, spotted here and there with white evening primroses and pink desert five-spots. The bursts of flowering are not necessarily correlated with the year's rainfall. For instance, the wettest year in Death Valley was 1941, when 4.2 inches of rain fell, but there was no mass flowering that year or the following spring. If Death Valley is to bloom in the spring, the rain must come at a certain time—during the preceding November or December. There will be a mass display of spring flowers if November or December has a precipitation of well over one inch: in December of 1938 and in November of 1946 the rainfall was 1.4 inches. Rain of this magnitude in August, September, January or February seems ineffective.

Let us consider these annual plants in greater detail. Probably their most remarkable feature is that they are perfectly normal plants, with no special adaptations to withstand drought. Yet they are not found outside the desert areas. The reason lies in the peculiar cautiousness of their seeds. In dry years the seeds lie dormant. This itself is not at all amazing; what is remarkable is that they refuse to germinate even after a rain unless the rainfall is at least half an inch, and preferably an inch or two. Since the upper inch of soil, where all the viable seeds lie, is as wet after a rain of a tenth of an inch as after one or two inches, their discrimination seems hard to explain. How can a completely dormant seed measure the rainfall? That it actually does so can easily be verified in the laboratory. If seed-containing desert soil is spread on pure sand and wet with a rain sprinkler, the seeds will not germinate until the equivalent of one inch of rain has fallen on them. Furthermore, the water must come from above; no germination takes place in a container where water only soaks up from below.

Of course this sounds highly implausible—how can the direction from which the water molecules approach make any difference to the seed? The answer seems to be that water leaching down through the soil dissolves seed inhibitors. Many seeds have water-soluble germination inhibitors in their covering. They cannot germinate until the inhibitors are removed. This can be done by leaching them in a slow stream of water percolating through the soil, which is what happens during a rainstorm. Water soaking up in the soil from below of course has no leaching action. Some seeds refuse to germinate when the soil contains any appreciable amount of salt. A heavy rain, leaching out the salts, permits them to sprout. Other seeds, including those of many grasses, delay germination for a few days after a rain and then sprout if the soil is still moist—which means that the rain probably was fairly heavy. Still other seeds have inhibitors that can be removed only by the action of bacteria, which requires prolonged moisture. Many seeds preserve their dormancy until they have been wet by a series of rains.

In the washes (dry rivers) of the desert we find a completely different vegetation with different germination requirements. The seeds of many shrubs that grow exclusively in washes (paloverde, ironwood, the smoke tree) have coats so hard that only a strong force can crack them. Seeds of the paloverde can be left in water for a year without a sign of germination; but the embryo grows out within a day if the seed coat is opened mechanically. In nature such seeds are opened by the grinding action of sand and gravel. A few days after a cloudburst has dragged mud and gravel over the bottom of a wash, the bottom is covered with seedlings. It is easy to show that this germination is due to the grinding action of the mudflow:

for instance, seedlings of the smoke tree spring up not under the parent shrub itself but about 150 to 300 feet downstream. That seems to be the critical distance: seeds deposited closer to the shrub have not been ground enough to open, and those farther downstream have been pulverized. Smoke-tree seedlings form about three leaves, then stop their above-ground growth until their roots have penetrated deep enough to provide an adequate supply of moisture for the plant. Thereafter the roots grow about five times as fast as the shoots. Few of these seedlings die of drought, but a flood will destroy most of them; only the oldest and biggest shrubs resist the terrific onslaught of rocks, gravel, sand and mud streaming down the wash.

The ability of the smoke tree to make the most of the available moisture was demonstrated by the following experiment. Cracked smoke-tree seeds were sown on top of an eight-foot-high cylinder containing sand moistened with a nutrient solution. Rain water was then sprinkled on them for a short time. Six seeds germinated, and five of the plants survived and have grown for 18 months in a high temperature with only a single watering midway in that period. Indeed, they have grown better than seedlings which were watered daily;

We have studied the control of germination in great detail in our laboratory at the California Institute of Technology. We have learned, for instance, that two successive rains of three tenths of an inch will cause germination provided they are given not longer than 48 hours apart. Rain in darkness has a different effect from rain during the day. Most amazing is the seeds' specific responses to temperature. When a mixture of raintreated seeds of various annuals is kept in a warm greenhouse, only the summer-germinating plants sprout, the seeds of the winter annuals remain dormant. When the same seed mixture is kept in a cool place, only the winter annuals germinate. From this it is obvious that the annuals will not germinate unless they can survive the temperatures following their germination—and unless there has been enough rain to allow them to complete their life cycle. Since these desert plants cannot depend on "follow-up" rains in nature, they germinate only if they have enough rain beforehand to give them a reasonable chance for survival.

A very small percentage of seeds (less than 1%) germinate after an insufficient rain. Such seedlings almost invariably perish before reaching the flowering stage. On the other hand, more than 50 per cent of all seedlings that have sprouted after a heavy rain survive, flower and set seed. And here we find a remarkable fact: even though the seedlings come up so thickly that there are several thousand per square yard, a majority of them grow to maturity. Though crowded and competing for water, nutrients and light, they do not kill one another off but merely fail to grow to normal size. In one case 3,000 mature plants were found where an estimated 5,000 seedlings had originally germinated. The 3,000 belonged to 10 different species. All had remained small, but each had at least one flower and produced at least one seed. This phenomenon is not peculiar to desert plants. In fields of wheat, rice and sugar cane, at spots where seeds happen to have been sown too thickly, all the seedlings grow up together; they may be spindly but they do not die. It is true that in gardens weeds often crowd out some of the desirable plants, but usually this happens only because these plants have been sown or planted out of season or in the wrong climate. Under those conditions they cannot compete with the plants fully adapted to the local growing conditions—plants which we usually call weeds.

We must conclude, then, that all we have read about the ruthless struggle for existence and the "survival of the fittest" in nature is not necessarily true. Among many plants, especially annuals, there is no struggle between individuals for precedence or survival. Once an annual has germinated, it matures and fulfils its density of forming new seed. In other words, after successful germination annual plants are less subject to the process of "natural selection." Very likely this accounts for the fact that so few of the desert annuals seem to show adaptations to the desert environment. This does not mean that the plants have avoided evolution, but the evolution has operated on their seeds and methods of germination rather than on the characteristics of the grown plants. Selection on the basis of germination has endowed the plants with a remarkable variety of mechanisms for germinating, and at the same time it has made them show to germinate except under conditions insuring their later survival. The opposite is true of the cultivated plants that man has developed: his selection has favoured the plants that germinate most easily and quickly. This has given us the wrong perspective on the significance of germination in plant survival.

We return now to our original theme: Can the ecology of plants in the desert teach us anything about human ecology or human relations? At least one moral stands out. In the desert, where want and hunger for water are the normal burden of all plants, we find no fierce competition for existence, with the strong crowding out the weak. On the contrary, the available possessions—space, light, water and food—are shared and shared alike by all. If there is not enough for all to grow tall and strong, then all remain smaller. This factual picture is very different from the time-honoured notion that nature's way is cut-throat competition among individuals.

Actually competition or warfare as the human species has developed it is rare in nature. Seldom do we find war between groups of individuals of the same species. There are predators, but almost always they prey on a different species; they do not practice cannibalism. The strangler fig in the tropical jungle, which kills other trees to reach the light, is a rare type [see "Strangler Trees," by Theodosius Dobzhansky and Joao Murca-Pires; Scientific American, January, 1954]. Even in the dense forest there is little killing of the small and weak. The forest giants among the trees do not kill the small fry under them. They hold back their development, and they prevent further germination. In a mountain forest in Java it was observed that the small trees living in the shade of the forest giants had not grown after 40 years, but they were still alive.

Hundreds of different species of trees, large and small, grow in a tropical jungle. This diversity of vegetation is one of the jungle's most typical characteristics. Some trees grow faster, taller or wider than others, but these growing characteristics, which we have always considered as useful adaptations in the struggle for existence, do not really control the trees' survival. If they did, we would find very few species of trees in a jungle, and there would be an evolutionary tendency for these trees to become taller and taller. Actually the tallest trees are found not in jungles but in more open forests in temperate climates; remarkably enough, tropical jungles often have no particularly high or large trees. All this shows that selection does not work on the basis of growth potential. It works on the ability of plants to grow and survive with very little light.

In our minds the struggle for existence is usually associated with a ruthless termination of the less well adapted by those better adapted—a sort of continuous cold war. There is no cold war or even aggression in the desert or jungle. Most plants are not equipped with mechanisms to combat others. All plants grow up together and share whatever light or water or nutrients are available. It is only when the supply of one of these factors becomes critical that competition starts. But it appears likely that in the jungle, as in the desert, survival is taken care of by the control of germination. Competition and selection occur during germination, and we can speak of germination control of the plant community—comparable to birth control in human society.

Apparently evolution has already eliminated most of the plant types that are unable to compete successfully. Fast-growing, slow-growing or tall plants all have the same chances once they have germinated. The struggle for existence is not waged among the well-established plant forms but tends to eliminate new types which germinate at inopportune times, have a decreased ability to photosynthesize or are less frost-resistant. This explains why so few plants die in the desert from drought or in the jungle from lack of light or in cold climates from frost.

As a general moral we conclude that war as man wages it finds no counterpart in nature, and it has no justification on the basis of evolution or natural selection. If we want to describe the process of control of the plant population in human terms, we should talk about birth control.

9

Community Dynamics

Introduction

The previous chapter stressed the fact that a community is a functional entity: this complexity of structure and organisation lacks a dimension, until it is seen in operation for it is a system which only exists in motion. Clearly the operation of any community is a function of the interaction of the various processes affecting its component organisms: competition; parasitism predation and so on, and these individual processes will each be examined in detail in later chapters. But to stick first with our 'overview' let us consider the operation of the community as a whole: the net result of all these separate interactions combined.

This is, of course, somewhat more easily said than done, and for many years ecologists lacked this overview. They had founded their science on meticulous observation and field recording; with such an approach, a picture of community function only could arise as the product of the interactions of its component species. It was not possible to study the dynamics of the whole, for suitable techniques were simply not available; even within the community it was often not possible to form more than a qualitative, subjective, estimate of the roles of the various organisms, and there were many problems that could not be tackled at all, or others which, through their complexity, could only be partially resolved. In short, both in terms of understanding relationships within the community, and in understanding the dynamics of the whole, ecologists were looking for a more rigorous unifying approach, a method of quantification with respect to some very basic common theme. And they took their cue from the theoretical physicists, by turning to a study of energy flux.

The Community as a System of Energy Transformations

It is indeed valid to consider any biological system—individual, population or community—as a system for the transfer, storage and dissipation of energy. It is, what is more, a system that obeys all the basic laws of thermodynamics: that energy can neither be created nor destroyed, and that each transfer, or transformation of energy cannot be 100 per cent efficient: energy will be lost and there will be an increase in entropy at each stage. Animals and plants during their lifetimes act as stores of potential energy. In feeding or in being fed upon, that energy is transferred to and from other organisms. Not all the energy is retained as potential energy within the body: the life processes of maintenance—basal metabolism, or whatever you like. require energy expenditure; still more energy is released in 'work'. Thus some of the original energy is dissipated from the system in respiration.

Within an ecological system, of course this 'energy' is not free energy but is the chemical energy incorporated into the structure of complex organic materials—the energy of molecular bonds. Animals, and certain heterotrophic plants (fungi and many bacteria) take in their energy ready-stored within the organic molecules of their food, but what is the ultimate source of energy for the ecological system? For this we must look back to the Producer level: to those autotrophic organisms that are able to synthesise organic compounds from inorganic precursors, using free energy in the process. Certain bacteria use energy derived from the low potential chemical energy of inorganic chemical bonds—in 'chemosynthetic' processes. The majority of autotrophs however are photosynthetic, using the energy input from solar radiation to form the bonds of their synthetic compounds. Thus, in one way or another 'inorganic' energy is taken from the environment and used to synthesise 'organic' energy, by the autotrophs—the producers. They use the energy thus derived both for growth and respiration. That used for growth may be taken in by the primary consumers, and passed on up the trophic scale. However, not all the consumed material is usable by the herbivore; some passes direct to the decomposers in faeces and urine. What is retained by the consumers (or assimilated) may then also be used for growth and respiration. If they in turn are then eaten by other consumers, the small amount of energy they have stored in tissue production is again diminished. At each stage the amount of energy passing to the trophic level above decreases and at each stage, energy is dissipated through respiration (for that, too, is the ultimate fate of that material which passes to the decomposers). Ultimately all the energy fixed by the producers is dissipated in respiratory activity, so that eventually all the solar (or chemical) energy entering the system is released in the heat energy of respiration—conforming with the first law of thermodynamics noted earlier: no energy is lost and none gained. (Note that while nutrients and basic elements cycle through ecosystems the flow of energy is strictly linear. Odum (1959) has coined a simple catchphrase as a mnemonic 'Matter circulates, energy dissipates'.)

So energy does flow in a logical way through biological systems, and this provides the unifying link which enables us to study the mechanics of the system or its operation: for we may analyse the flow of energy through the system' quantifying the efficiency of transfer from organism to organism, the speed and direction of that transformation and the use to which energy is put within the organism or trophic level, and we can compare communities with respect to these various parameters. As always we have again the opportunity of looking at the community as a jigsaw of its component species, or in viewing it as a single functional whole. Here we may study the energy relationships of individual organisms, and by linking them into their appropriate context, build up a picture of the community, or we may look at the general flow of energy through the community in its own right.

Energy Relationships of Individuals

The energy relationships of an individual organism (or species) may be defined in terms of its Consumption, and its Assimilation (the amount of energy retained after excretory losses—in animals, faeces, urine, gases of fermentation). Thus

Assimilation = Consumption − Faeces and Urine

or, (adopting the conventional terminology of Petrusewicz, 1967):

$$A = C - Fu \qquad \ldots. (9.1)$$

Similarly, the use to which assimilated energy is put may be represented. Assimilated energy may be used for Production (P) in both individual growth (Pg) and reproduction (Pr), and for respiratory costs (R) of basic maintenance (R_m) and work (R_w). Thus

$$A = P + R \qquad \ldots. (9.2)$$

$$\text{or } A = Pg + Pr + Rm + Rw \qquad \ldots. (9.3)$$

All these individual quantities may be precisely determined in terms of calories, or joules (kgm^2s^{-2}).

Analysis of these values and their relationships may tell us a great deal about the operation of the organism itself. For example, the relationship of Assimilation to consumption (Assimilation efficiency: A/C per cent) reflects efficiency of food use: the assimilation efficiency of most herbivores is low—of the order of 30-40 per cent—while that of carnivores is in general much higher (63-80%). (The 'assimilation efficiency' of plants is surprisingly low. Green plants trap only 1-5% of the sun's energy in photosynthesis.) A derivative of this relationship is the agricultural Production efficiency (P/C%). Analysis of the way in which the assimilated energy is apportioned between Production and Respiration (P/R) is also very illuminating in informing us whether or not this is a very active organism which in consequence has little energy to spare for production, or whether it needs little energy for respiration and puts most of its available resources into growth. This last is a relationship that changes rapidly through the life of an individual, too: younger animals and plants tend to be very fast growing and (since respiratory cost is related to body size) have relatively low respiratory expenditure; older animals have higher respiratory costs and less energy to set aside for growth and reproduction. We can learn a lot in detail about the use of the 'Production' energy: whether it is used in growth, purely in size increase, growth in replacement of insect larval cuticular exuviae, pupal cases etc. which are discarded, growth in reproduction. All this information can be quantified to draw interesting conclusions about the organism's use of energy and energy balances. From such knowledge we can even make predictions about its ecology. If we know the energy costs of foraging for a certain type of food, and the energy return (assimilation) from feeding on one such food item, we can make predictions about the way in which an animal may forage, its prey choice and pattern of hunting, whether it will accept or reject any given items and so on. We can even make suggestions about the predators ideal size, and whether or not the species may show sexual dimorphism (Schoener, 1969). Further, the information can be used in an applied sense, too. From calculations of daily energy consumption and assimilation, knowledge of tempiratory costs of maintenance and activity, we can estimate the net production surplus any individual animal will retain after satisfying respiratory costs from a measured assimilation. If we also know the energy costs of reproduction for that species we can predict, from a mere survey of food availability, whether or not that individual or species will be able to breed in any given season a prediction which may be of great value if the species concerned is an economic post.

Energy Relationships in the Community

From knowledge of the energy relations of an individual, we can, as we have noted, build up a picture of the energy relations of the whole community. For the Consumption of an individual or species must be taken from somewhere else in the community, while Production (if the system is in steady state) must be used either to sustain standing crop biomass in replacing those individuals which have dieds or to supply food for other organisms in the system. Dead material, faeces and urine are inputs for the 'Consumption' of the decomposers; respiratory releases are the only loss. So for each organism we can refine the simple

$$C—Fu=P+R$$

to fit it into the community framework. If we repeat the exercise the full pattern of energy flow through the community well emerge. Such analyses have been made (*e.g.* Teal, 1959; Odum, 1959) but are perhaps of limited value—except insofar as a detailed analysis is required of some particular system for some specific purpose for they permit of little gentralisation. A detailed analysis is offered of one specific system, and is true only for that system; further, because of the complexity of interactions within a community, it is necessarily only very simple systems, or parts of systems, which can be treated in this way (and it notable that the classic studies of Teal and Odum were both carried out on small freshwater springs). In addition, even within such restricted systems the complexity of relationships is such that it is not possible to draw any conclusions which might be applicable to communities in general. To do this we must adopt the same rationalisation as we invoked in the previous chapter; regarding the community as a system of discrete trophic levels. This approach—and indeed the whole concept of the community as a series of trophic classes—was first developed by Lindeman in 1942 in a study of. . . a freshwater spring!

Energy Flow Within the Community: the Tropho-Dynamic Approach

This reduction of the community structure to a series of trophic levels makes analysis much more manageable. We may then examine various aspects of the passage of energy through the system: amounts of energy transferred, the speed of transfer through the system, and the efficiency with which energy passes from one trophic level to the next; we may also consider the use to which the energy is put within each trophic level.

Energy Relationships within a Trophic Level

Within any one trophic level, of course, the use and apportionment of energy can be examined by exactly the same analysis as applied to evaluation of the energy relationships of an individual organism or species population. Thus we may summarise the relationships in terms of the equations

$$C-Fu=A=P+R$$

In the same way, we can define trophic level assimilation and production efficiencies (A/C%, P/C%). Relationship of the production efficiencies of various trophic levels within the same, or different communities make interesting comparisons (*e.g.* Table 9.1). Other analyses of exactly the same type as those carried out for individual organisms or species population, offer figures for the amount of material held in living tissue within the trophic level—in standing crop biomass, rate of turnover and so on: all measures defining the structure and dynamics of the trophic class. But it is in the transfer of energy to and between the different trophic levels that the real keys to community function can be discovered.

Energetic Efficiency

One of the major considerations here, as we have already hinted, is of the efficiency with which energy is transferred through the ecological system. Not all the energy any animal

TABLE 9.1

Net Production Efficiency (Net Primary Production/Gross Primary Production) of Several Plants and Plant Communities

Plant community	*Locality**	*Net production efficiency (per cent)*
Terrestrial		
Perennial grass and herb	Michigan	85
Corn	Ohio	77
Alfalfa		62
Oak and pine forest	New York	45
Tropical grasslands		55
Humid tropical forest		30
Aquatic		
Duckweed	Minnesota	85
Algae	Minnesota	79
Bottom plants	Wisconsin	76
Phytoplankton	Wisconsin	75
Sargasso Sea	Tropical Atlantic Ocean	47
Silver Springs	Florida	42

Source: From Odum, 1959.

takes in its food is retained as potential energy in growth; much is lost in assimilation, and of that assimilated energy, more is expended in respiration. Nor is the entire production of a population of animals or a trophic level, passed on to the trophic level above; much is unavailable to the consumers and many individuals will die unconsumed. Of that which is consumed, again only a proportion will be assimilated. Thus if we look at efficiency of energy transfer between trophic levels as

$$\frac{\text{Energy consumed by trophic level n}}{\text{Energy consumed by trophic level below (n—1)}} \quad (9.4)$$

we will always find it considerably less than one. Indeed, in terms of percentage, this 'gross ecological efficiency' is usually somewhere between 7 and 14 per cent. It has been suggested that the efficiency is in fact a constant: that the range of figures derived between 7 and 14 per cent is due to necessary inaccuracy in measurement and that in natural ecosystems, the efficiency of energy transfer between trophic levels approximates to a constant 10 per cent: an axiom always attributed to Slobodkin (1961). But, while this may be a useful rule of thumb, it is in fact the very differences in efficiency which are most interesting. Efficiency of energy transfer may differ between different pairs of trophic levels within one community—offering an insight into the different mechanics and efficiency of operation of different trophic classes, to complement knowledge derived from analyses of energy flux within the trophic levels themselves. In the same way, efficiency of transfer may also differ between equivalent pairs of trophic levels in different communities, hinting at differences in methods of operation between the various communities (Table 9.2.) Many of these) differences in operation and differences in energy flow we shall explain in later chapters.

While gross ecological efficiency (or Lindeman's efficiency) is the ratio must commonly referred to in studies of community function, other relationships are also invoked. To avoid confusion in reading other literature these are summarised together with efficiencies relating to individuals or populations and thus by extension to trophic levels in Table 9.3.

The depletion of energy availability up the trophic scale is frequently quoted, in explanation of Eltonian pyramids of numbers or biomass. If only a small percentage of the energy entering one trophic level is effectively transferable to the level

TABLE 9.2

Efficiency of Energy Transfer at Various Trophic Levels in Three Aquatic Ecosystems

	Trophic Level Energy Intake Efficiency (per cent)		
Trophic level	*Cedar Bog Lake, Minnesota*	*Lake Mendota, Wisconsin*	*Silver Springs, Florida*
Photosynthetic plants (producers)	0.10	0.40	1.2
Herbivores (primary consumers)	13.3	8.7	16
Small carnivores (secondary consumers)	22.3	5.5	11
Large carnivores (tertiary consumers)	not present	13.0	6

Source: From Odum, 1959.

above it, then the biomass supported must be reduced. The same argument is then extended as explanation for the limit to the number of trophic levels themselves. As Pimm and Lawton (1971) have recently observed, however, this last claim is not easily reconciled with the fact that the number of trophic levels appears to be a constant—and essentially independent of enormous variations in the amount of energy flow and transfer efficiencies: 'food chains are not noticeably shorter in barren Arctic and Antarctic terrestrial ecosystems compared with a productive tropical savannah, or the fish guilds of a tropical coral reef. Energy attenuation obvious ultimately provides an upper limit to the number of possible trophic levels in any given system. Pimm and Lawton suggest that the actual number of trophic levels observed is well below this limit, and that the actual limit to the number of trophic levels observed (and hence the apparent constancy) is determined in the dynamics of the various populations of the community. They use mathematical studies of the stability properties of variously constructed model food webs to argue that long food chains may typically result in population fluctuations so severe that it becomes impossible for higher order predators to persist. This suggestion is further developed by Saunders (1978). Vincent and Anderson (1979) and Pimm and Lawton (1978).

TABLE 9.3

Various Types of Ecological Efficiency (based on Odum, 1959)

Ratio	*Explanation*
A. Ratios within trophic levels	
$\frac{P}{A}$	Tissue growth efficiency
$\frac{P}{C}$	Production efficiency (agriculturally: food conversion efficiency)
$\frac{A}{C}$	Assimilation efficiency
B. Ratios between trophic levels	
$\frac{C_t}{A_{t-1}}$	Gross ecological efficiency or Linderman's efficiency (Energy intake by trophic level t/Energy intake of trophic level t—1)
$\frac{A_t}{A_{t-1}}$	Trophic level assimilation efficiency
$\frac{P_t}{P_{t-1}}$	trophic level production efficiency
$\frac{A_t}{P_{t-1}}$	Utilisation efficiency

Speed and Amount of Energy Transfer

Other facets of the flow of energy through the community, besides pure efficiency, can provide useful pointers to important aspects of community function. Thus the speed with which energy passes through the system differs between different communities, whether measured as speed of turnover within one trophic level or through the whole community; and this may have important implications for community structure and stability. De Angelis et. al., (1978) have shown that relatively unstable model food webs can be kept effectively stable provided that the transit time for a unit of energy through that web is relatively fast.

TABLE 9.4

Average Transit Time of Energy in Living Plant Biomass (Biomass/Net Primary Production) for Representative Ecosystems

System	*Net primary production* ($g\ m^{-2}\ yr^{-1}$)	*Biomass* ($g\ m^{-2}$)	*Transit time* (*yrs*)
Tropical rain forest	2,000	45,000	22.5
Temperate deciduous forest	1,200	30,000	25.0
Boreal forest	800	20,000	25.0
Temperate grassland	500	1,500	3.0
Desert scrub	70	700	10.0
Swamp and marsh	2,500	15,000	6.0
Lake and stream	500	20	0.4 (15 days)
Algal beds and reefs	2,000	2,000	1.0
Open ocean	125	3	0.024 (9 days)

Source: From data in Whittaker and Likens. 1973 atfer Ricklefts, 1971.

Further the amount of energy passed through the system at any one time (defined by input, in terms of energy fixed by the producers and then passed on by them through the community) may also differ between communities. The total amount of material synthesised within the community by the producer level is referred to as the Gross Primary Production (GPP). That which is available to other elements of the community after respiratory expenditure by the producers (GPP—Respiratory cost) is defined as Net Primary Production (or NPP). That GPP—and thus the amount of energy passing through the whole community—can vary enormously between different systems is illustrated by the summary table taken from Odum (1959), while Odum himself has shown that the quantity and quality of energy flow through a community are indeed tremendously important features of that community, defining indeed its whole structure and mode of function.

TABLE 9.5

Gross Primary Productivity of Various Ecosystems as Determined by Gas Exchange Measurements of Intact Systems in Nature

Ecosystem	*Rate of production ($g\ m^{-2}\ day^{-1}$)*
Infertile open ocean, Sargasso Sea	0.5
Shallow, inshore waters, Long Island Sound; year average	3.2
Texas estuaries, Laguna Madre	4.4
Clear, deep (oligotrophic) Lake, Wisconsin	0.7
Shallow (eutrophic) Lake, Japan	2.1
Bog lake, Cedar Bog Lake, Minnesota (phytoplankton only)	0.3
Lake Erie, Winter	1.0
Lake Erie, Summer	9.0
Silver Springs, Florida	17.5
Coral reefs, average three Pacific reefs	18.2

Source: From Odum, 1959.

Limitations of Energy Analysis

To review: the study of the flow of energy through an ecological system enables us (1) to put the individual organisms of the community in context, and to define more exactly their actual function in the system, (2) to extend this if we wish to a detailed analysis of the entire flow of energy through specific systems and, (3) to restrict the consideration to one of energy flow between trophic levels and apportionment within them.

However, in pursuing our analysis of community function purely in terms of the transfer and transformation of energy we must bear in mind the possible limitations of this approach. For consideration of the community merely as a system for the transformation of energy is of course a gross oversimplification. It restricts consideration to trophic relationships alone (all others are neglected) and further presupposes that energy is the single most important commodity passing through the system. In reality, while of course organisms need energy in order to function, they require other raw materials as well—nitrogen to build protein, potassium, phosphorus, calcium—a whole host of other nutrients. And in practice, at least in the producer and primary consumer levels of the community (which are, after all, the two major elements within the structure) it is very unusual to find energy in itself to be limiting. Far more likely to be in limited supply are these other nutrients, particularly nitrogen and potassium. Energy, as we have seen, has a linear flow through any ecological system; nutrients cycle and since the available 'pool' is limited in extent, bottlenecks within the system—in say, release of nitrogen from dung or corrion—are likely to have considerably greater effect in restricting community function than energy itself. While analysis of energy flow is conceptually simple and straightforward, and so to speak 'embodies' the total operation and functions of an organism, population or trophic level, in a single, quantifiable measure of impact, it should always be regarded as a first approximation only. It is not 'sufficient up to itself'. Working on the energy dynamics of a species of grassland bug (Notostira sp.), Gibson (1977) found that, at the time when females began to prepare and develop eggs and at just that time when one might anticipate they should feed where maximum energy intake would be assured, they moved from a grass of high calorific value to one of considerably lower energy content: a grass in fact, upon which they could only barely satisfy their minimum energy requirements. In terms of energy alone: a paradox, resolved only when it is appreciated that the second grass is extraordinarily high in nitrogen. In another, similar study, Greenstone (1979) has shown that wolf-spiders (Pardosa ramulosa) feeding on three main prey species (the shorefly Ephydra riparia, the mosquito Aedes dorsalis and waterbatman Trichocorixa reticulata) seem to select prey with regard to aminoacid composition rather than energy, selecting a diet that optimises their own intake of eessntial dietary aminoacids. The importance of nutrients other than energy has also been recognised in recent studies by, for example, Westoby (1974), Pulliam (1975b), White (1978) and Stenseth and Hansson (1979). Nonetheless, studies of energy flux within communities give a good first approximation, a good first impression of dynamics and methods of function—information which could not be obtained by any other analysis. As a complementary approach to many ecological problems, consideration of energy flow will crop up in a number of later chapters, together with information derived from a more classical, observationally based approach, to supplement and clarify many of the points we will discuss.

The Flow of Nutrients Within Communities

As we have already noted, the flow of energy through any ecological system is linear, and relies on continuous new input at the producer level. By contrast most other materials cycle through the ecological system: there exists a 'finite' and specified amount of each nutrient within the system which cannot easily be replenisned; in this case continued function of the system relies on the coutinuous cycling of each nutrient within it.

Maintenance of life depends on a whole host of different nutrients: water, carbon, oxygen, nitrogen, potassium, phosphorus, calcium, iron, magnesium, sulphur—and a variety of other minerals required only in minute quantities (so-called trace elements). Many of these nutrients are superabundant, but supplies of others are potentially limiting. It is the dynamics of these nutrients—required in large amounts in proportion to their overall availability—with which we are primarily concerned here: these are the nutrients likely to restrict community function; these are the nutrients for which efficiency of transfer and circulation is most critical.

'*Nutrient Pools*': *Free Nutrients and Biota*

We may consider, in essence, a limited quantity of each such material available within the system: a limited nutrient 'pool'. Some proportion of this may be immediately available as 'free' nutrients in the soil, ready to be taken in by the primary producers of the system as required, while the remainder of the ecosystem's supply of each nutrient will actually be bound up temporarily within the living tissues of its component organisms: so to speak, currently in use. The relative size of the 'free nutrient' pool and that proportion bound up in the biota of the ecosystem are important considerations. If there is a relatively large stock to available nutrients then the continued operation of the system is not immediately limited by the speed of return of 'used' nutrients for re-use. It can, so to speak, live off its fat, and continue to operate efficiently by utilising the large stock of free-nutrients which will ultimately be replenished by the return of those nutrients presently 'in currency' when they are finished with. Nutrients do not move through living systems in smooth even flow, but in pulses, jerks and floods. The larger the stockpile or free nutrients, the greater the capacity of the system to accommodate the vagaries of return of those nutrients bound up within the tissues of living organisms. If, by contrast, very little of the available pool of a particular nutrient is 'free' within the system, that system becomes effectively limited in its operation by the speed of return and recycling of materials from the living component.

In general, in relatively simple and undeveloped ecosystems, only a small proportion of the available nutrients are incorporated in the tissues of living organisms at any one time. As the community develops however, and becomes more complex, the biomass of living organisms supported increases dramatically; more and more of the total 'nutrient pool' becomes trapped in the biota. In these more complex communities, the efficiency and speed of nutrient turnover indeed becomes limiting; in such communities the importance of the decomposers in releasing such nutrients from the dead bodies and excretory wastes of animals and plants cannot be overemphasised.

10

Aircraft and Airport Noise

Aircraft have been probably the most dramatic of the man-made noise sources which are heard by the general community, especially in the vicinity of an airport. Although the number and power of aircraft increased gradually in the 1940s and 1950s, it was not unit the introduction of jet-powered engines that their potential noise nuisance became widely recognised. In many advanced countries the major objective surveys of aircraft noise levels around airports and companion social surveys of the effects of noise on nearby residents have been undertaken.

There are many thousand people living around. Airports, who are moderately or seriously annoyed by airport noise. No estimate is available of the numbers of people affected by aircraft noise around major and minor airports and airforce bases. In most surveys of noise annoyance, aircraft noise runs a close second to traffic noise in terms of the number of people affected and the extent to which they are annoyed.

As the purpose of an aircraft engine has been to take in air and then generate a thrust by throwing out air behind it, and as noise is generated by the interaction between an ejected airstream and the surrounding atmosphere, it is not surprising that an aircraft engine is noisy. However, as the mechanical energy which appears as sound has been only a very small fraction of the total energy produced by the engine, it is very difficult to quieten. Despite this, most of the major airport authorities and the airlines take the problem of aircraft noise very seriously. Their interest gets accelerated since noise certification in advanced countries whereby every new aircraft of an existing type manufactured after 1976 and every aircraft of a completely new type manufactured after 1969, must satisfy the requirements of a quality control test with regard to noise emission.

In order to begin to discuss how to keep out aircraft noise from a home, or how to influence the take-off procedure and routing of aircraft so as to alter the noise nuisance or similar problems, it is necessary to look at the units which measure aircraft noise. Subsequently, methods of control of aircraft noise including insulation, aircraft and airport operating procedures, planning controls and aircraft design have been examined.

This source of noise pollution is increasing steadily during recent years and, especially close to international airports, already constitutes a very serious problem. This problem has mainly arisen because of the widespread use of heavy long-range jet aircraft. Noise made by jet planes has been intrinsically more disturbing than that of propeller-driven aircraft because it is of far higher pitch. Jet noise has been caused by the violent mixing of the jet of gases from the engine with the surrounding air; it is at a maximum during take-off when the engine must deliver maximum thrust, and falls away rapidly as the aircraft climbs. During landing, the main source of high-frequency noise has been the whine of the air compressor and the turbine blades as the engine is throttled back. Aircraft pass close to the ground for quite a distance during the landing operation and this noise often constitutes a more sustained environmental nuisance than the instense noise of shorter duration produced during take off.

Military aircraft often make annoyance in areas away from airfields because they are to be flown at low altitudes as part of normal training procedures. Little can be done about this; national deference—even is peace-time—will always take priority.

The internal wreckage caused by shriek of siren or the road of a jet engine has been including gastric ulcers and thymus gland atrophy. While the sonic boom (shock waves produced by an object moving faster than the speed of sound) at our threshold, mankind gets threatened with the most awesome noise pollution. According to Prof. Garret J. Hardin, "Sonic boom has been much worse than noise. Experiencing it is like living inside a drum beaten by an idiot at insane intervals."

WHAT TO DO ABOUT AIRCRAFT NOISE

Once again, the primary advice to those who are sensitive to aircraft noise has been stay away or move away from noisy areas. Those who have a natural tolerance or sense of identification with aircraft may live in the affected area. Having said that however, there will be those who for many other reasons have to live in affected areas, or people who regard that aircraft noise is an imposition that should get reduced so as to allow them to live again in peace. Again, some may wish to move away but will not feel it fair to stand personally a loss in cash due to reduced house value because of the noise, or who cannot afford to move due to high mortgage commitments which the reduced house value may not cover. All these people have to seek means to combat the noise.

FINAL REMARK ON THE METHODS OF REDUCING NOISE FROM AIRCRAFT

To reduce noise from jet aircraft during take-off, noise suppressors can be fitted—but these also reduce take-off thrust and increase fuel consumption, drag and weight. The additional operating cost caused by fitting noise suppressors to an aircraft of the size of a Boeing 707 is about £ 16500 per annum. Newer types of jet engine, in which the jet velocity is lower, show a reduction of as much as 12 dB against older engine designs, and with these there is little point in fitting noise suppressors. Modern take-off procedures over densely populated areas are also planned to reduce noise. Full power has been used to unable the aircraft to climb as rapidly as possible before it reaches a built-up area: power than gets reduced until the built-up area has been cleared. But such procedures mean that aircraft loadink has to be reduced, either by taking on less fuel, fewer passengers or less baggage. This, naturally, tends to increase operating costs. So far it has been found impossible to do much about landing noise, although research in this field is proceeding.

There has been increased commercial pressure to extend night operations so as to improve the profitability of aircraft and airports. Although there is mounting public resistance against this trend, the number of night flights from most airports continues to show a steady increase.

It appears to the authors that many airlines and most civil aviation authorities have been doing as much as possible to reduce the noise nuisance from aircraft; further major improvements can only come from the re-siting of airports away from centres of population, preferably so that both take-off and landing flight-paths have been over the sea. For the present, the only way to avoid the detrimental effect of noise originating at large airports is to provide individual sound-proofing for houses, hospitals and schools in their vicinity. Openable double-glazed windows, complete with a mechanical ventilator and equipped with a noise attenuator, can reduce aircraft noise by about 40 dB, as against a figure of 20 dB with normal single windows.

THE SUPERSONIC TRANSPORT AND SONIC BOOM

Whenever a solid body travels at a speed in excess of the speed of sound, a sonic boom gets produced which can be heard up to 80 km from the point of origin. The actual pressure exerted has been found to vary with the distance above the ground. A supersonic passenger aircraft flying at a height of 12,000 m can produce a pressure jump at ground level of up to 100 N/m^2 (127 dB) which may not bring about actual damage to buildings, but can still be an unpleasant experience for many people. Supersonic passenger aircraft also bring about other side effects including the destruction of ozone in the outer atmosphere, with consequent increase of harmful radiation. The construction and sale of supersonic aircraft has been at present a matter of national prestige and economics; but from the environmental standpoint these aircrafts are most undesirable.

The supersonic transport (SST) brought noise pollution problems. A supersonic airplane is one that travels faster than sound. When an aircraft travels at a velocity greater than the speed of sound, there is an enormous increase in air resistance. A shock wave is created that can be extremely energetic, depending of the size, speed, and route of the aircraft. When a shock wave strikes things below, it can have a jolting and devastating effect. The shock wave, or sonic boom, may spread

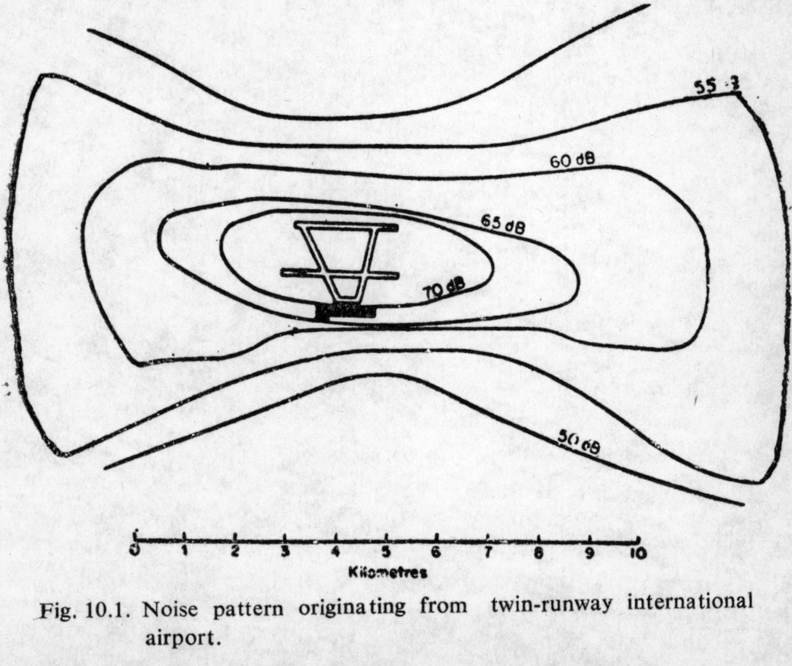

Fig. 10.1. Noise pattern originating from twin-runway international airport.

over an area of 10 to 80 miles. Although many cases of damage to structures caused by sonic booms from relatively small military aircraft have been reported and numerous claims damage have been filed, the destructiveness and nuisance effects of large commercial supersonic aircraft have not been fully evaluated. Objections to the development and commercialization of an SST are based on the possibility that their shock waves might become intolerable.

Noise pattern originating from twin-runaway international airport is shown in Fig. 10.1.

11

Effects of Noise on People

The transmission of sound through the ear and the mechanism of hearing are described briefly in this section. The ear can be considered in three parts which are generally called the outer, middle, and inner ears. The car canal (auditory meatus) extends from the pinna to the ear-drum (tympanic membrane). The canal itself is approximately oval in crosssection and extends for about 25 mm into the head.

The middle ear is an air-filled cavity which contains a chain of three small movable bones called the auditory ossicles (the malleus, which is partly within the ear-drum, the incus, and the stapes), as well as two minute muscles which contract in an involuntary reflex response to intense sounds. The original function of this reflect contraction, which reduces the transmission of sound across the middle ear, is not clear but it thought to be related to the improvement of speech discrimination rather than the protection of the inner ear from loud noise. The Eustachian tube connects the middle ear cavity to the pharynx and allows the equalisation of air pressure on either side of the ear-drum.

The footplate of the steps is located in the oval window leading to the inner ear. The inner ear is a complex system of fluid-filled cavities positioned deep in the temporal bone of the skull so that it is physically well protected. The system of cavities includes the cochlea and three semicircular canals oriented at right angles to each other. These canals from part of the body's balancing mechanism and need not concern us further.

The cochlea transduces the mechanical vibrations of the ossicies into neural impulses. It is extremely small (about 8 mm in diameter), filled with fluid, and similar in shape to a snail's shell, having about 2¾ turns coiled round a central bony pillar which also carries the auditory nerve fibers from the cochlea to the brain.

The cochlea itself is split longitudinally into two fluid-filled sections by the cochlear partition which is itself a tube filled with another fluid. The sections of the cochlea are clearly shown. The lower part of the partition is formed by a flexible membrane ealled the basilar membrane which supports the hair cells. There are three rows of outer hair cells and one row of inner hair cells. They are the sensors which initiate neural impulses to the brain to produce the sensation of hearing. Immediately above the hair cells is the tectorial membrane which is relatively rigid, and the hairs of the hair cells are probably attached to its undersides. The upper part of the cochlear partition is called Reissner's membrane.

It becomes necessary to have an understanding of how the individual parts of the ear behave so that when we come to discuss hearing loss and deafness we can appreciate what is happening.

Sound first enters the ear through the outer canal. It is a small tube about 25 mm long and varying in diameter. Little physical damage can be done to this tube except that it does tend to collect low grade bacteria and once a colony of these is established it is very hard to shift. The sound then fails on the eardrum, which is not flat but is a cone with an included angle of about 120°; the apex is on the inside of the eardrum. The eardrum vibrates under the influence of the incident sound wave. This vibration then gets connected through a series of bones which reduce the amplitude but increase the force upon inner 'windows' which, in turn, transmit the waves to liquid borne pressure waves within the circular canal called the cochlea. The pressure fluctuations within the liquid in the cochlea then excite small nerve cells which are known as hair cells. Each of these hair cells has its own individual nerve and these nerves are connected together into the auditory nerve which then goes off to the appropriate brain centre.

It has been important to remember that the human brain has been a component of this system. The brain works as an on-line computer interpreting the signals it receives from all the hair cells in the cochlea. There has been a lot of parallel information coming in along the various nerves and it is a mistake to think for example that an individual nerve cell corresponds to a particular frequency. In fact, groups of nerve cells together tend to be responsible for human rural response at particular frequencies. Due to the complicated nature of this system it can be seen that the human ear is likely to react in different ways at different frequences of incoming sound. In fact, the ear tends to be most sensitive at about 3000 Hz because of a resonance in the outer ear canal. The sensitivity of the ear falls off rapidly at high and low frequencies and a graph of the relative sensitivity. It can be seen that complicated effects set in when the intensity of the sound varies, so that whereas at low intensity the frequency response of the ear varies quite markedly, at high intensities (about 90 to 900 dB) the subjective response of the human ear has been, for all intents and purposes, flat (that is, independent of frequency).

ADVERSE EFFECTS OF NOISE

The generation of unreasonable noise within the environment is regarded as a form of pollution because it lowers the quality of life. There has been several specific ways in which excessive noisc can affect people adversely.

Noise has been found to interfere with our activities at three levels:

(a) audiological level in refering with the satisfactory performance of the hearing mechanism;

(b) biological level interfering with the biological functioning of the body; and

(c) behavioural level affecting the sociological behaviour of the subjects.

Because of this, the noise affects stating categorically, performance, physiology and psychology.

Noxious has been known to cause of nervous disorder, headache, high blood pressure and short memory. The various effects of noise may be outlined as follows:

NOISE EFFECTS ON HUMAN BEINGS

Noise hazards
Permanent hearing loss
Neural-humoral strees response
Destruction of artifacts

Noise nuisance

Efficiency	Comfort	Enjoyment
Mental stress	Sleep interference	Concentration interference
Frustration	Communication	Meditation interference
Task interference	Invasion of privacy	Recreational interference
Irritability	Damage of artifacts	Temporary hearing loss
	Habit of talking louy	

Nose contributes to development of cardiovascular problems like heart disease and high blood pressure. Workers exposed to high noise level are having more circulatory problems, cardiac disturbances, neurosensory and motor impairment and even more social conflicts at home and at work.

TABLE 11.1

Effects of High Intensity Noise on Human Beings

Noise (dB)	*Effects observed*
0	Threshold of audibility.
150	Significant change in pulse rate.
110	Stimulation of reception in skin.
120	Pain threshold.
130—135	Nausea, vomiting dizziness, interference with touch and muscle sense.
140	Pain in ear, prolonged exposure causing insanity.
140	Extreme limit of human noise tolerance.
150	Prolonged exposure causing burning of the skin.
160	Minor permanent damage if prolonged.
190	Major permanent damage in a short time.

1. Damage to the Hearing Mechanism

Occupational hearing loss is not a new phenomenon; for example 'boilemaker's deafness' was well known among foundry workers in the nineteenth century, and gunpowder explosions are reported to have caused deafness amongst gun-crews in the Battle of Trafalgar.

The ear can be damaged in a number of ways. Physical poking of objects into the outer canal can rupture the eardrum. This is usually not completely fatal to the nearing mechanism and eardrums can be repaired. A very large shock wave can sometimes physically break the bones of the middle ear which transmit the sound across from the air to the lymphatic liquid in the cochlea and when this happens instaneous deafness can occur, but again this can often be repaired by skillful surgery and prostheses. Once the sound waves have entered the cochlea they are then picked up by the hair cells and it is these hair cells that are damaged by prolonged exposure to high intensity sounds. It is best to think of these hair cells as slow blow fuses. They can withstand a short period of high intensity sound but then start to deteriorate and are broken. It is an unfortunate fact that in general nerve cells within the human body do not regenerate themselves. So that once a hair cell has been damaged it is totally lost.

Medical experts now belive that our hearing worsens with age partly because of exposures to the typical noise of our industrialized society. Dr. Samuel Rosen, ear surgeon from Mount Sinai School of Medicine in New York, studied the Mabaans, a primitive African tribe living in quiet environment. He found that Mabaan man of 75 can hear as much as the average 25 years old American man.

Over a period of a lifetime progressive deafness occurs as these cells die off. This is called presbycusis. Many people think that presbycusis is a natural phenomenon, but others think that it is partly a result of the high noise levels that we have in our society. In fact, many audiologists (scientists concerned with the ear and hearing) have adopted another term, sociocusis, to represent the loss of hearing caused by general, non-occupational, noise exposure. It seems probable that people who live in urban areas have less sensitive hearing than those who live in remote rural areas and who experience little or no industrial noise. Indeed, the self-inflicted noises of many of our leisure pursuits—such as discotheques—are making sociocusis an increasing reality.

It is now well understood that excessive noise can cause damage to the ear: in fact, any continuous noise level above about 85 dB (A) will cause some damage. This damage is slow, insidious and progressive and is not usually noticed by the recipients until it is far too late. A lot of people who have been working in high noise level industries all their lives are now quite deaf in their old age. This is a subject which is becoming increasingly of interest in many countries because it has now been established that employers can be held liable for deafness caused to their employees merely by noise alone.

Exposure to intense noise may lead to a hearing loss, which will appear as a shift in the hearing threshold. Some of the hearing loss which results from exposure to intense noise is usually temporary, and over a period of time, possibly as long as several weeks, partial or complete recovery will occur. Such temporary loss is usually referred to as temporary threshold shift (TTS). Any hearing loss which persists is called permanent threshold shift or noise induced permanent threshold shift (NIPTS). Permanent threshold shift is associated with the destruction of the hair cells in the cochlea. Hence, once a permanent threshold shift has occurred, normal hearing cannot be restored.

In some recent work, Goran Bredberg performed postmortem examinations on 119 cochleas of 75 patiente. He found that the most hair cells were present in the fetal cochleas and that there was a progressive loss of hair cells from birth on. At age twenty years, about 90 per cent of the hair cells remained, and at thirty years, approximately 85 per cent of the cells remained. Furthermore, the exact percentage varied with the location in the cochlea. Hearing sensitivity for low frequency sounds related best to the condition of the hair cells in the apical region of the cochlea, while sensitivity for high frequency sound retated best to the condition of the cells near the base of the cochlea. However, while a loss of 20 per cent of the hair cells in the apical region of the cochlea resulted in no well-defined loss of hearing sensitivity, the loss of 10 to 20 per cent of the cells in the basal region of the cochlea seemed to result in a hearing loss of up to 40 dB for high frequency sounds. This region also appears to be more susceptible to damage because resonances in the outer ear canal apparently amplify the intensity of sounds with frequencies near 4,000 Hz. These two factors are apparently responsible for the rather sharp drop in hearing sensitivity at 4,000 Hz frequently observed in people who have been exposed to noise. Noise-induced hearing loss is a function of three main factors:

1. overall noise level,
2. the frequency compositions of the noise, and
3. the total duration of exposure.

Even moderate levels of industrial noise can therefore cause hearing loss if the individual is exposed to such noise eight hours a day for a period of many years.

Noise interferes with speech, sometimes making it unintelligible; sounds of warning are misunderstood or not heard. Even low-level noises impair verbal communication because they require more attention and effort and cause ministerpretations. Noises produce irritability and a feeling of fatigue and may reduce a worker's efficiency. There is evidence that noise is one of the major causes of stress and many of the other human afflictions associated with tension—anxiety, insomnia, accident-proneness, high blood pressure, and other cardiovascular diseases. Even in its less serious forms, noise is commonly a source of annoyance and cause for complaint. All these are of personal, social, and economic importance.

The effects of noise on the fetus are not fully known. Medical scientists have noted that an unborn child will move and kick when there is a loud noise. It also respounds with a sudden increase in the heartbeat as through it were disturbed or frightened.

The effect of environment noise on the foetal development during pregnancy has been two subject of research. It has been reported that constant exposure to noise between 110-120 decibels can produce narrowing of vision, vertigo and disruption of equilibrium in the urborn baby.

It has been found that startling can quicken a human fetus's heart rate and cause its muscles to contract. Noise may cause malformation in fetus's nervous systems.

Many workers who are exposed to the noise of jet aircraft or very noisy workshops for even moderate periods soon develop detectable hearing defects. These days it has become the practice for workers in these situations to be equipped with ear protectors and—provided these are worn all the time—hearing ability has been not affected.

In big cities like Bombay, Calcutta and Delhi, the average noise level has been found to be between 65 and 90 decibels. The noise level doubles every six years, taking its poll of air hearing. By 2000 A.D. it is possible that no one above age ten will hear normally.

2. Repeated Interference with Sleep

In a social survey carried out amongst people living in the vicinity of Palam Airport, some 22 per cent said that they sometimes found difficulty in getting to sleep because of airport noises. In areas where the noise level was particularly high, up to 50 per cent complained about the noise. An even higher percentage said they got awarkened by high intensity noises usually early at night when sleep was not yet deep. After people have been asleep for some hours they do not readily wake up, even when subjected to very loud noises. Different people have different depths of sleep and they can adjust to nocturnal sounds. Undoubtedly, however, noisy conditions near residential areas as night must get avoided; lack of continuous sleep is having a very harmful physiological effect on human beings. If it is impossible to reduce external noises, it becomes necessary to insulate dwellings effectively to counteract it. Many techniques for sound insulation are available today and can be applied at relatively modest expense.

Noise reduces the depth and quality of sleep and thus may adversely affect overall mental and physical health. Dr. Cohn Heridge a British psychiatrist, found in a two year study that persons living alongside London's Heathrow Air port have a significantly higher rate of admission to mental hospitals than persons living in socially similar but quieters areas.

A sound need not be loud enough to completely arouse one;

but may change the level of sleep. For instance, interruption of the REM (Rapid Eye Movement) stage where dreams occur, has a curious effect. The person has to compensate later and feels very fatigued.

3. Effect on Communication

External sounds are able to interfere with conversation and use of the telephone, as well as the enjoyment of radio and television programmes and like pastimes. It can thus affect the efficiency of offices, schools and other places where communication has been of vital importance. The maximum accepted level of noise under such conditions has been 55 dB. 70 dB is considered very noisy and serious interference with verbal communications is inevitable.

4. Effects on Mental and Physical Health, and Working Efficiency

Many people complain that noise makes them mentally or physically ill and reduces their working efficiency. Experiments have been carried out to attempt to confirm or disprove these claims. According to the H.M. Stationery Office report *Noise*, noise certainly does not contribute in the least to mental illness. Damage to the inner ear may occur if continuous noise levels in excess of about 100 dB are experienced, but there is little evidence that it can lead to physical illness.

Noise is also having little effect on the performance of either physical or mental tasks, provided that the level is below about 90 dB. There are many people who complain they are unable to work effectively if a clock is ticking too loudly or if there is whispered conversation nearby; in these cases the fault lies in the subject's inability to concentrate, rather than in tha physical or physiological effect of noise. Some people—including the author of this book—find it easier to concentrate under conditions of slight noise than in situations where there are no background sounds at all.

Intermittent noises may bring about distraction and could

therefore affect the efficiency of people who have to give constant attention to a single task. Efficiency does suffer above a sound level of 90 dB, and experiments have shown that far more errors are made than at lower sound levels. This has been found to be true even when the subject has been working for a long time at high sound levels and should be accustomed to the noise.

5. Physiological Responses

Sound may produce a number of reflex-like reactions. The most common of these is the startle response which is usually caused by loud, unexpected sounds and is evident as a flexing of various muscles and a blinking of the eyes. These reactions in themselves do not have any known effect on health, although in many work situations they could startle a person into injuring himself or others. Physiological changes accompanying the startle response and other noise exposures include:

(i) A vascular response characterised by peripheral vasoconstriction, changes in heart rate and blood pressure.

(ii) Various glandular changes (such as increased output of adrenalin) evidenced as chemical changes in blood and urine.

(iii) Slow, deep breathing.

(iv) A change in the electrical resistance of the skin with changes in activity of the sweat glands.

(v) Brief changes in skeletal muscle tension.

There is, however, no direct evidence that these changes are associated with any harmful effects. Instead, a variety of inderect evidence is often cited: for instance exposure of rats and rabbits to a mixture of sounds with average level of 83 dB has been found to cause enlarged ventricles and increased heart weight relative to groups of non-exposed control animals over a period of three weeks. Extrapolation of animal studies to humans is, however, brought with uncertainties.

By contrast, laboratory studies with people are constrained by the relatively low noise levels and short durations which can be used. The majority of tests reported are of short duration, so it remains open to question whether the body adapts to noise with prolonged exposure.

In an evolutionary context these physiological responses appear to have a survival value—an increase in blood supply to the brain and improved grip due to moistening of the palms being of value during an emergency. Whilst noise is clearly a stressor in evoking these changes, it need not necessarily have harmful effects. An increased output of adrenalin during a period performing a simulated industrial task in 'realistic industrial noise' may cause no greater harm than similar increases when viewing films depicting violence. Alternatively, it is possible that repeated evoking of these changes with regular prolonged exposure to noise may contribute to health disorders. Evidence for such long-term effects could be provided by field studies which are considered next.

Several birds have been observed to have stopped laying eggs. Apart from this the animals change their places. There is a decrease in migratory birds to a place if it is subject to noise. Prolonged chronic noises can also produce stomach ulcers as it may reduce the flow of gastric juice and change its acidity.

Some empirical research conducted on animals (especially pregnant female mice) which are not possible to be conducted on human beings on ethical and moral grounds reveal that aircrafts taking off which brings in 120 to 150 decibels caused miscarriages in them. If the findings hold generalizability, high noise is capable to create these disturbances in human beings also.

According to environmentalist Thomas G. Aylesworth, "*Constant noise may cause our blood-vessels to contract, our skin to become a pale, our muscles to construct and adrenaline to be short into our blood-stream.*" This adrenaline is responsible for both—excitatory and inhibitory—responses in living beings. This is the reason that factory workers develop abnormal heartbeat rates and suffering from insomania, nervousness and impaired motor coordination. Considering these points, U.S. Government has set *90 decibels as a health hazard* for an 8 hour day working environment.

It has been experimentally proved that high noise is bad particularly for those suffering from high tension, diabetics. The roadside noise may not only wake up a person but also alter his heart beat and respiration. Noise also produces startling effects on babies and they may even develop a fear psychosis as a result of sudden and sharp noise.

Soviet studies show that workers exposed to high noise levels had twice the incidence of hypertension and upto four times the incidence of peptic ulcers.

6. Epidemiological Studies

Several researchers have conducted field studies testing industrial workers and/or collating their health records in an attempt to overcome the limitations of duration and realism in laboratory studies. A classic study by Jansen examined the occurrence of physiological problems amongst two groups of workers he classified as from "very noisy industries" and "less noisy industries" respectively. The differences between the two groups, indicate a higher incidence of problems amongst the high-noise group than in the low-noise group. Noise level, however, is not isolated in such a study as the *causative* factor. The two groups were in fact based on two separate industries, the high-noise group being steel-workers whereas the low-noise group came from lighter industries. Between these industries there would be numerous other differences which could have effects on health, such as heat, physical work load, anxiety, and the type of people first attracted to the respective industries. At most, therefore, one can conclude that noise may have been a *contributory* factor in the differences found.

A study by Cohen partly overcome this problem by locating different areas within the same boiler manufacturing complex. Comparing information from medical records of two noise groups, indicated that the high-noise group had a substantially higher proportion of diagnosed disorders.

In a continuation of this study Cohen reported an apparent reduction in the medical problems of the high-noise group after the introduction of hearing protectors, thus providing some further support for the hypothesis that the noise levels were at least a contributory factor to the differences observed.

Using measures of blood pressure, Parvizpoor reported three-to-four times as many hypertensive amongst a group of weavers than in a control group working in 'light' less noisy industries. By contrast a similar study by Malchaire and Muller did not reveal any significant differences between two groups of noise-exposed workers (categorised by degree of hearing loss) and a control group who were not exposed to noise. Conflicting results are also found in studies which employed measures of hearing loss as direct estimates of the noise exposure.

The limited number of studies in this orea prevents the drawing of any firm conclusions as to whether noise is, or is not, associated with adverse health effects. It would appear from these and other studies that noise may be evident (if at all) as an additional stressor. For instance in the study by Jansen noise appeared to cause an additional 10 per cent of peripheral circulation problems on an already high (50%) occurrence rate amongst the low-noise group. Noise may therefore have an effect by increasing the probability of the appearance of symptoms amongst susceptible individuals.

7. Performance Effect

Another major area of concern is the effect that noise may have on task performance. It is likely that any new sound or change in an existing sound may result in at least momentary distraction, and that this may impair a person's ability to perform some tasks. More generally, changes *via* the non-specific connections in the cortex associated with prolonged exposure to even relatively steady noise may have some effect on such task performance. Extensive research both through field and laboratory studies has failed, to date, to produce definite criteria; however, recent research has brought to light the mechanisms whereby noise may interfere with task performance.

It has been observed that children's performance is poor in 'comprehension' tasks whose schools are situated in busy areas of a city and suffer from noise pollution. Noise, because of lacking rhythm, causes irritation which results in learning disabilitles. Sudden noise distracts a person and can crete nervousness in him. House-wives working in kitchen with all kinds of electric gadgets have been seen to get headache due to noise and vibrations of these gadgets. People working in printing presses are prone to such effects and become deaf due to such noise.

A classic early study by Weston a Adams found that loom operators working in a 96 dB noise environment had a greater output when wearing earplugs. Those individuals who were most aware of a distracting influence of noise showed the greatest improvement with earplugs, but an improvement was evident even amongst those who believed themselves to be unaffected by noise. In spite of this, Broadbent argues that the differences observed were probably due to suggestion and a morale boost to the weavers.

Many other such studies have been plagued by this 'Hawthorne Effect,' named after a series of studies at the GEC works in Hawthorne, Chicago, where it was found that any change of work conditions, be it an improvement or a degradation, resulted in increased production. The experiment was confounded by the improved motivation of the workers due to the interest shown in them by the experimenters.

Similarly a study by Broadbent and Little found that there was an increased rate of working by operators of film-perforating machines when working in rooms where the noise level had been reduced from approximately 98 to 99 dB. However, the same workers were also studied in rooms which had not been acoustically treated, and similar improvements were observed. This was attributed to improved morale of the workforce, an effect which appeared whichever room they were working in at the time. In the same study it was observed that the number of broken rolls of film attributable to faults of the operators was significantly reduced by the noise treatment compared with the change in the untreated rooms. The results provide some evidence that noise reduction reduced human error due to momentary lapses of attention, whilst it did not improve work rate, except perhaps by a general morale boost. These findings indicate that our understanding of noise effects will require consideration of details of the type of task and the measure of task performance used.

A large number of laboratory-based experiments have investigated under controlled conditions the effects of noise on task performance. A review of these experiments is provided by Broadbent.

Recent research has indicated that noise at levels as low as 80 dB(C) may significantly impair task performance relative to quiet conditions. More generally, effects are reported for noise levels greater than approximately 90 dB(C), Larger effects have been observed for noise with predominant high-frequency components, and for time-varying noise.

Tasks most prone to disturbance are those that have little margin for error, require constant attention over long periods, involve complex mental manipulations, or require interacting with more than one source or sensory channel. Typically, the impairment in performance takes the form of signals being missed, increased errors in response, and prolonged response times.

The results of these studies tend to be highly specific to the particular experimental conditions used. They have, however, brought to light the underlying mechanism of the effects of noise.

The effects of noise on task performance are seen to be complex, depending on the nature of the task and the noise, and the level of arousal of the individual. The evidence, however, is clear that levels of noise above approximately 80 dB(A) can impair performance at tasks which have little margin for error and require continuous attention, and that there can be associated 'psychic costs' which become manifest both during and after performance of the task. The difference between laboratory conditions and typical work situations prevents the existing data being reliably applied to individual work situations.

8. Effects on Auditory Skills

Hearing plays an important role not only in speech communication but in the perception of a variety of important sounds. It provides information about our environment, and as the 'sentinel of our senses' often acts as the first warning of impending danger. It can allow identification of the source of the sound, give information about the location of the source, and in a more subtle manner the degree of reverberation gives a 'feel' for the acoustic environment. Noise can impair of eliminate perception of this information.

A recent review has indicated that whilst it has often been said that noise can be the cause of accidents, there is relatively little data available to quantity such as an effect. More-generally, a recent World Health Organization report noted that "communication requirements in industrial situations frequently do not receive adequate attention, particularly with reference to the accident risk." Studies such as those reported by Cohen have indicated higher accident rates in high-noise areas of factories. Whilst results such as these do not establish a causal link, there are several ways in which noise could give rise to such an effect. Firstly, the 'psychic costs' discussed in the preceding section associated with working in noise could make a person more likely to be involved in an accident by increasing fatigue and impairing the perception of unexpected hazards. Other mechanisms could be associated with the masking effect of the noise, which could impair the perception of warning sounds, auditory cues, and speech messages. These effects are considered next.

9. Speech Interference

Speech reception is the most important and also the most complex use of the auditory system. Noise can either mask speech to make it inaudible, or by masking only some frequencies leave it audible but of reduced integlligibility. The interfering effect of noise on face-to-face speech communication can be assessed in terms of the maximum possible distance between the speaker and the listener start in a given steady background noise level for a particular voice level. For personal conversation where a separation of 2 m is typical it can be seen that normal communications are possible in noise levels up to approximately 60 dB(A). At higher levels the talker will unconsiously raise his voice level to compensate for the high-noise environment, and with this greater effort communications may still be just reliable in noise levels up to approximately 70-75 dB(A).

It still higher noise levels it would be neccssary to move closer together if possible, or to provide some aid to speech communication. For discussions involving groups, such as during training and instruction, separations of 4 m may be more typical, and in these cases noise levels should not exceed approximately 55 dB(A)

The exact spectrum of the noise is also an important factor, and for more accurate assessment of a noise environment, reference should be made to either the Articulation Index (AI) or the Preferred Speech Interference Level (PSIL). More general assessment of indoor noise levels are provided by the Preferred Noise Criteria (PNC) curves or by Noise Rating (NR) curves.

There are important implications of the reduced opportunity for relaxed, reliable speech communication. In a noisy industrial environment, for instance, it will be considerably more difficult to instruct new employees on the job. Whilst conditions may be adequate for the predictable information associated with some jobs, important unexpected messages may not be comprehended which could be a hazard to safety and the efficient operation of plant equipment. It has been suggested that as workers get used to the noise environment they adopt a non-communicating life style, relying more or non-verbal communication through gestures and facial expression, which impairs the development of social relations and degrades their work environment.

10. Personological Effects Produced by Sound

If the injurious effects of noise tend to persist for long, they cause stable maladaptive reactions in the individual disturbing his total personality make up. The lowered performance level among children may develop a feeling of inadequacy, lack of confidence, poor perception of one's ownself which may jeopardize their optimal personological development as a growing child. Our lack of awareness of the causative factors of these maladaptive developments in child may encourage us to follow those remedial lines which may further aggravate the problems instead of resolving it. Once the feeling of ineptness, worthlessness and inadequacy are developed by a child in the formative period, its disasterous effects are not going to be removed easily without leaving their marks behind.

11. Effects of Noise of Non-Living Thing

The high intensity of noise effects of non-living things too, *viz.*, buildings. The cracks in the city of the Recoco Church at Stemhausan were caused by some booms.

12. Noise Pollution and Number of Accidents

Some scientists of U.S.A. have discovered that high pitched noise leads to road accidents. In India also this correction resulted in number of fatal accidents as follows:

TABLE 11.2

Number of Fatal Accidents in Different Cities of India

Year	*Number of fatal accidents*		
	Delhi	*Bombay*	*Calcutta*
1980	720	918	1156
1981	1021	1375	1732
1982	1172	1951	2156
1983	1559	2384	2983
1984	1956	2979	3413

12

Wildlife Conservation

Introduction

Wildlife Conservation includes all human efforts to perserve wild animals from extinction. It involves the protection and wise management of wild species and their environment. Some species have become extinct due to natural causes, but the greatest danger to wildlife results from human activities. Thus, we ourselves have created this need for wildlife conservation.

The progress of man throughout has been beneficial for the human race but it is the wildlife that has suffered through the years. Invention of sophisticated weapons, industrialization, urbanization, ever increasing human population have been some of the major causes for the dwindle of our once rich resource. Hunting, clearing of forests, draining of swamps and damming of rivers for irrigation and industry this is what we appraise of man's progress. These activities have vastly reduced the natural habitats of our wildlife and many species are endangered or nearly extinct.

Man's indifference and ignorance had led him to believe that he is independent of natural laws and of natural environments and forgets that he himself is a biological being and is directly and/or indirectly dependent on the natural resources available to him. He exploits, destorys and reduces the renewable natural resources, eradicates wildlife and leads his own race to levels leading to catastrophe. All this can be put to an end by careful, long-range ecological planning.

Importance of Conservation

Alongwith the topic of conservation arises a question—Why should be care about disappearing animals? The answer to such a question can be attributed to many reasons. Some of them are as follows.

Survival Value. Every species plays an important role in maintaining an ecological balance among the living systems of the earth. These systems must continue to function if life is to survive. Loss of any species threatens the survival of several species inclusive of man. Man's own survival depends on his willingness and ability to co-operate with the living environment. Human beings turn to nature and wildlife for spiritual, pscholo-gical, physical and cultural benefits.

Scientific Value. The study of wildlife provides valuable knowledge about life processes which has helped scientists understand the functioning and behaviour of the human body. Scientists have also gained medical knowledge and discovered important medical products by studying wildlife. By studying the effects of environmental pollution on wildlife it is possible to learn how pollution affects human life.

Economic Value. Wild species of animals provide meat for food and skin for fur. The financial value of wild species is of major importance in the economics of many nations.

Beauty. Wildlife includes a diversity of species that contri-butes in a special way to the beauty of nature. Such beauty enriches life and increases the enjoyment of camping and out-door recreation.

On all continents wildlife is becoming an increasingly important recreational asset and tourist attraction. They maintain the health of the living landscape and productivity of the eco-system.

Wildlife represents the result of many years of evolution and constitute an environmental heritage to the past, a world which once surrounded our ancestors. Moreover, why should man deny that animals as species have the right to live and coexist with ourselves? We are responsible to the coming generation to leave as an inheritance an unspoiled world with intact living beauty and richness.

Apart from the aesthetical and ethical reasons there are also educational arguments to preserve wildlife

Reasons for Dwindle

Man in his zeal for progress fails to realize that his own existence is being jeopardised. "Progress", as he terms it, has made him contaminate air; pollute streams, lakes and oceans, drain swamps and wetlands, lowering the water table; rape prairies and forests; spray pesticides over water and land; kill animals excessively for immediate profit, or on false assumptions that they are noxious or for fashionable "sport". Man alone, though often in combination with his livestock, is destroying the natural environment.

"Civilization", as we apply the term, is not just cities, industrial developments, miles of highways, enormous airfields, and numerous other technological conveniences. By destroying wildlife for civilization and culture, we are only negating the meaning of the words.

If he does have to change habits, then he must be able to manage them properly in order to maintain a rich and varied wildlife. Man has to think of the dangerous trop he is setting up for himself human overpopulation.

It is man's ability to think that formulated the idea of conservation is applied ecology and may also be regarded as bioeconomy. In other words, conservation is defined as sound preservation, management and sensible use of the available renewable natural sources. This decision of conservation is in our hands and man seems to be going in the right direction regarding this decision that holds the key to the future of our most valuable natural resource. Governments of several nations have passed laws and set aside national parks, sanctuaries and other wildlife reserves in an effort to save the dwindling wildlife. Such efforts have saved several species from becoming extinct. However, these efforts have to increase several fold, if not for a complete recovery, at least about 3/4 of the species threatened with extinction can be redeemed.

Classification of Scarce Wildlife

There are three major categories of wildlife that are facing extinction.

Endangered Species. They face the most severe threat of extinction and require direct human protection for survival.

Threatened Species. They are abundant in certain areas but face serious dangers nevertheless. Dangers being caused due to unfavourable environmental conditions, extensive hunting, fishing, trapping or even collecting by hobbyists.

Rare Species. They have small populations. They live in protected environments and their numbers are not decreasing.

Methods of Wildlife Conservation

This depend on the source of danger to the threatened species. In many cases, wildlife can be helped by ensuring that their environment is favourable with regards to availability of food, water and shelter. This method is called Habitat Management. It involves such action as soil conservation, good forestry pr actices and wildlife management.

Several species have been threatened with extinction due to destruction of habitat by human activities like deforestation for farming, pollution of air, water, land, plants and animals etc. People must set aside seperate areas like sanctuaries, wildlife reserves, parks etc. In which wildlife can survive. Hunting is another major cause for the dwindle in the wildlife and this can be prevented by enforcement of strict laws that forbid or regulate such killings.

Excessive increase in the number of one species in a given area threatens not only its own endurance but also that of the other species by consuming major portion of the available food. Such numbers must be reduced by controlled hunting or by restoring its natural enemies where they have become scarce.

Sometimes species find it extremely difficult to survive in their own once favourable environment. Such species can be protected by breeding in captivity and releasing in a protected area that satisfies the conditions favourable for survival.

A species threatened with disease can be protected by sanitation measures in a habitat.

The success of wildlife conservation depends on the knowledge of the *Ecology* of the species which involves on uaderstanding of the way of life of the species and its relationship with its biotic and abiotic environment.

History of Wildlife Conservation

The story of the struggle for wildlife conservation is several decades old. If all started with few foresighted men who understood the role of wild animals in living landscapes and the necessity of preserving the threatened species. Encouraged by the pioneering efforts of these men several national and international organizations were formed for conservation of the most valuable of our natural resources.

The first of such organizations was formed by a group of

Ornithologists in 1922, the International Council for Bird Preservation (ICBP). The first post-War organization to be formed under the sponsorship of UNESCO was the International Union For Conservation of Nature and Natural Resources (IUCN) in 1948. In 1961 the IUCN formed the first fundraising organization, the International Wildfowl Research Bureau (IWRB). There are several other organizations like the Food and Agricultural Organization (FAO) of the United Nations, the Smithosian Institute and the Conservation Foundation both in Washington DC and the Fauna Preservation Society in London.

The Zoologists Gessellschaft in Frankfurt, West Germany, has raised important funds for conservation, particularly for Africa.

The IUCN though its Survival Service Commission (SSC), is dealing specifically with animals threatened with extinction. Their function is to collect data on endangered species throughout the world and to initiate action to prevent the extinction of these species. The SSC investigates the status and ecology of a species and advises governments and organizations. Such advice is often continued with direct action financed by WWF. The SSC has long maintained a list of threatened mammals, birds, reptiles amphibians, fishes etc., called "The Red Data Book".

This conservation expansion has produced positive results in many parts of the world. Inspite of such efforts, the future remains uncertain for the world's wildlife. This uncertainity exists and intensifies with the increase in human population, destruction of wildlife habitats and spread of environmental pollution.

Wildlife Conservation in India

The Indian Wildlife possesses a rich and diverse fauna and is unique in having immense natural beauty in its different ecosystems. The fauna includes about 123 families of terrestrial vertebrates. An estimate puts the number of 400 mammals, 1200 species of birds, 350 species of reptiles and more than 29,70,000 species of insects.

The animals that make up the Indian fauna are Elephant, Rhinoceros, Musk deer (Kastura), Barking deer (Kotra), Spotted dear (Cheetal), Hog deer (Hoghiran), Mouse deer (Indian chevrotain), Swamp deer (Bara singha), Dancing deer (Sambur) Thamin, Kashmir stag. Four-homed antelope (Chausingha), Indian Gazelle (Chinkara), Blue bull (Nilgai), Bison, Wild buffalo, Himalayan ibex or wild goat (Capra siberica), Wild boar, Wild ass, Nilgiri tahr, Lion, Tiger, Leopard, Striped hyena, Nilgiri langur (Ceropithecus John), Lion tailed macaque (Macaca silenus), Rhesus money (Macaca mulatha), Hanuman monkey (macaca mulath). Hanuman monkey (Semnopithecus entellus), Indian giant squirrel (Ratufa indica), Porcupine, Pangolin, Peafowl (Pavo cristatus), Junglefowl, Patridge, Quail Great Indian bustard (Choriotes nigriceps), Duck pigeon, Sand grouse, Stroks and Egrets, Pelicans, Eagle, crane, Owl, Hombill, Crocodiles, Gharials, Lizards (Uromastic) and about 216 species of snakes.

Endangered Species of India

The IUCN maintains a Red List of threatened species. This list is updated annually. A 1986 list contained the following species that were threatened with extinction.

These species face the most serious threat of extinction and require direct human protection for survival.

In order to protect these species from extinction we have to know about their habits and habitats. Once we are thorough with such facts, preservation of the endangerd species can be carried out successfully. The above table shows the list of endangered species in India and a detailed description of each animal is as follows.

TABLE 12.1

Order	Family	Zoological Name	Common Name	Status	Habitat
1	2	3	4	5	6
	Class Mammalia				
Primates	Cercopithecidae	Macaca silenus	Lion tailed macaque	E	Southern Western Ghats
Lagomorpha	Leporidae	Caprolagus hispidus	Hispid Hare	E	Assam
Cetacea	Platanistidae	Platanista indi	Indus dolphin	E	India
Carnivora	Viverridae	Viverra Megaspila civettina	Malabar spotted civet	E	Southern Western Ghats
	Felidae	Panthera leo persica	Asiatic lion	E	Gir Forest
		Panthera tigris	Tiger	E	India
		Panthera uncia	Snow leopard	E	Himalayas
Proboscidae	Elephantidae	Elephas maximum	Asiatic elephant	E	South NE India

Perissodactyla	Equidae	Equus henionus	Indian wild ass	E	Gujarat
	Rhinoceridae	khur Rhinoceros unicornis	Great Indian Rhino	E	Assam North Bengal
Artiodactyla	Suidae	Sus sulvanius	Pygmy hog	E	NE India
	Cervidae	Cervus duvaucelli	Hungal	E	Terai Kashmir
		Cervus elaphus	Swamp deer	E	Central India
		hungal Cervus eldi eldi	Manipur deer	E	Manipur
	Bovidae	Bubalis bubalis	Wild asiatic Water buffalo	E	Assam Bastar
	Class Aves				
Galliformes	Phasianidae	Tragopan caboti Tragopan	Cabot's Tragopan	E	NE India
		melano cephalus	Western Tragopan	E	Western Himalayas
Gruiformass	Otididae	Ardeotis nigriceps	Great Indian Bustard	E	Rajasthan MP
	Class Reptilia				
Testudines	Crocodylidae	Crocodylus porosus	Estuarine crocodile	E	East coast

(*Contd*).

1	2	3	4	5	6
	Garialidae	GA rialis gangeticus	Gharial	E	N Indian river
	Class Insecta				
Anoplura	Haematopinidae	Haematopinus oliveri	Pygmy hog sucking louse	E	Assam

E-Endangered species.

Macaca Silenus

Common Name—Lion Tailed Macque.

Local Name —Hind Siah Bundar, Malayalam—Nellamanthi, Tamil—Karungkorungoo, Arakkan.

Size —Head and Body-20-24 inches. Tail-10-15 inches (25-28 cms.) Females slightly smaller.

Distribution —Westerm Ghats from North Kamara, southwards to Kerala and Kanyakumari district, Tamilnadu.

Distinctive Characters —Great mane of long dark-grey or brownish grey hair from temple and cheek. Glossy black coat.

Habits —In habitats lonelier evergreen tropical forest belt between 2000-3500 ft. (610-1070 m). They are dark coloured, shy and seclusive in habit. They are seldom seen. Macaques are gregarious. They live in herds of 12-20 animals or more. The call of the male resembles that of the human voice—"COYEH" of the "COO" of a pigeon. Their movement through the treetops are slow and deliberate. Newborns are seen in September in South India.

Destruction of habitat and poaching have made this species an endangered species. Trees are the means of transport and form the habitat of this species. Hence afforestation and prevention of poaching are the methods to be adapted for saving this species.

Caprolagus Hispidus

Common Name—Hispid Hare of Assam Rabbit. Not much is known about this species except that they are the same size of the hare with a shorter, dark-brown tail. The for is coarse and is made of dark-brown. Bristles which are paler on the under-surface of the body. The outer side of the ear is brown. The eyes are small. Hindlegs are short and stout. Their claws are long and teeth are large. They feed on bark and roots. Usually seen in Terai, Dooars and Himalayan foothills. They are very elusive.

Panthera Tigris

Common Name—Tiger

Local Names —Hindi—Bagh, Central India—Nahar, Tamil-Poolee, Malayalam Koduwa

Size —Average length-9 ft. 9.6 ft. (275-290 cm). Female-8 ft. 6 in (260 cms.) Average weight—400-500 lbs (180-230 kgs). Female—100 labs less (45 kgs).

Distinctive characters —Tigers have a rich-coloured, well striped, short coat. Individual variations are great.

Distribution —They are found throughout India from the Himalayas to cape Comarin except in the deserts of Punjab, Sind, and Cutch. Also found in Burma.

Habits —Tigers are found at 10,000 ft. in the Himalayas, humid evergreen forests, in dry open jungle, grassy swamps of Terai, in Sunderbans leading an amphibions life, on trees, in mud and in water. Food, shade and shelter are essential for their survival. Tigers hunt from sunset to dawn, when it is not too hot. Its prey comprises of elephant (female and young), nilgai, deer, wildpig, bears, procupine, gaur, buffalo, panthers, other tigers and when extremely hungry feeds on fowl or fish. When game is not available, tigers attack cattle and even man. They climb trees and are excellent swimmers. Though they are heavy, they are athletic.

Mating is usually after the rains Young ones are born between February and May Male-female relationship ends when young are born: Gestation period is about 15-16 weeks, Tigresses give birth to 2 to 3 cubs per litter. The cubs begin to hunt when they are 6 months old and they stay with the tigress for 2 years. Sexual maturity is attained at the age of 3 years in the case of the tigress and 4 years in the case of the male.

Life span of a tiger is about 20 years. Hunting had greatly reduced the number in India. In 1970, a project was set in motion to prevent the degradation of the tiger population. This project was called "PROJECT TIGER". Nine sanctuaries were set aside as Tiger reserves for rehabilitation of the species. At the beginning of the project the number of tigers in India was less than 2000.

Panthera leo Persica

Common Name—Asiatic Lion.

Local Names —Hindi Sher, Gujarat—Untia bagh Malayalam Simham, Tamil—Singham.

Distinctive Characters —Mane is lesser than of the Africa Lion, cost is fuller and longer tassel of hair at the end of its tail, more, pronounced tuft of hair at the elbow joints and fuller fringe of hairs on its belly. The mane is light to dark, rarely, black. Cubs are spotted or striped.

Distribution —Once found in the whole of Northwest and Central India as far south as Namada. Now lions are restricted to the Gir forest Kathiwar, Gujarat.

Habits —Gir forest in Junagadh district covers about

500 square miles of rugged country and is composed of mainly stunted teak trees, palas, junbal, ber, patches of bamboo with an undergrowth of thorny shrubs and bushes.

Lions hunt at dusk and rest during the day. Their roars are heard throughout the night. The prey on game and cattle.

Lions do not have a particular breeding season. They are known to mate during October and November. Young ones are born during January and February. Gestation period is about 116 days. The male stays with the family and helps to defend the young and to get food for the young. Young ones are produced at intervals of at least 18 months to 2 years. A lioness gives birth to its first litter when it is 21/2-3 years old. There may be 2, 3 or even upto 5 cubs per litter.

In 1978, the number of lions in the Gir forest was less then 200.

Panthera Uncia

Common Name—Snow Leopard

Local Names —Simla—Barhal he, Kunawar—Thurwagh, Kumaon—Burhel haye.

Size —Smaller than a panther with a relatively longer tail.
Head and Body-3ft—3 in-3 ft 8 in (100-110 cm). Tail-3 ft (90 in).

Distinctive Characters —The snow leopard has a short muzzle, high forehead with a vertical chin. The colour of the coat is soft grey palling to pure white on the underside. Grey is sometimes tinged with buff. Spots are unbroken and distinct on the head, nape and lower parts of the limbs. On the body they break up into larger, paler rosettes. These rosettes are less pronounced in the luxuriant winter coat. The undersides are pale white except for a few black biotches and fur. Newly born cubs are darker than the adults.

Distribution —In India, the range extends along the whole of the Himalayan chain from Kashmir to Bhutan northwards into Tibet, Central Asia and Atlais.

Habits —Not much is known about the snow leopard. Its home is the higher altitudes of Himalayas at heights of 12,000-13,000 ft. (3660-3965 m) above sea-level, above the treeline where the rock and cliffs form the mountain sides. Lying by day, snow leopards hunt at night preying on wild sheep and goats, musk deer, hares, marmots and other rodents, or even larger birds. In summer, when the upland pastures are open to grazing, they follow their prey to these higher reaches and finding opportunity feed on domestic animals like goats, sheep or ponies. They establish territory near human settlements and feed on domestic animals. In winter, they move down with the animals to about 600 ft. (1830 m). The gestation period is about 3 months. Snow leopards give from 2-4 young per litter.

Uncias are persistently hunted for their valuable fur. Their soft colour and luxuriant beauty is scarcely rivalled.

Platanista Indi

Common Name—Indus Dolphin.

The indus dolphin was considered a separate species. Now considered conspecific with Platanista gangetic. They are found in the Ganges, Brahmaputra, Indus and other large tributories to the bases of hills. They are seen within tidal limits, but do not enter the sea. Its body oil is used for burning lamps and for various other purposes. Not much is known about this species.

Viverra Megaspila

Common Name—Malabar Large Spotted Civet.

This species was once very common in the coastal districts of Malabar and Travancore. It lied in the wooded plains and in the adjoining hill slopes. At present they are scarcely seen and appears to be nearly extinct. Nothing else is known about this species.

Elephas Maximus

Common Name	—Asian Elephant
Local Names	—Sanskrit Hasti, Gaja, Hindi—Hathi (male, Hathni) (Female), Marathi—Hatti Tamil—Anai, Malayalam Ana, Kannada—Ane
Size	—Average height-Adult male-9 ft. (2.75 m). Female—A foot lower (30 cms).
Distinctive Characters	—The Asian elephant is smaller than the African elephant. While the ears of the African elephant to enormous, that of the Asian elephant are smaller. The major distinctive feature is that of the hollow back of the African elephant to that of the rounded back in the Asian counterpart. Generally only the males have tusks, while females have a protrusion of a few inches or nothing at all.
Distribution	—In India, the Asian elephant is found in the Western Ghats, southwards from Mysore, Orissa, Bihar, Himalayan range, almost all over India.
Habits	—The Asian elephant is found in areas covered

with forests. They are extremely adaptable and live in steamy, humid jungle or in cool elevated forests. They live in herds of 5-60 animals (usually single families). When fodder is scarce herds break up into smaller parties. Big tuskers feed away from the herd. Bulls lead a solitary life and joins a herd when he feels the urge to mate. Elephants sleep when it is hot and feed in the mornings and evenings and during nights invade open forests and raid crops. They raid and retire after midnight. They feed on various kinds of grasses and leaves, stem and leaves of wild bamboo and plantain, all species of crops, barks of some kinds of trees. A full grown elephant eats about 600-700 lbs (270-320 kgs) of fodder a day. The exact breeding season is uncertain, but is estimated at the beginning of the rains. Young appear in Autumn. The gestation period is about 20 months. Elephants give birth to only one calf at a time. Elephants have poor sight but their sense of hearing and smell are well developed.

Hunting and poaching for ivory has very greatly decreased the number of elephants in India.

Equus Hemionus Khur

Common Name —Indian Wild Ass.

Local Names —Hindi—Gharkar.

Size —Height at shoulder-3 ft. 8 in (110 cms) to 4 ft. (120 cms)

Distinctive Characters —The colour of the coat varies from reddish grey to fawn or Pale chestnut. The erect dark brown mane is continued as a dark brown stripe extending along the back to the root of the tail, the lower being white.

Distribution —The Indian Wild Ass is found in the deserts of the Rann of Cutch and Peninsular India. The present population is less than 2000.

Habits —They roam in groups of 10-20, or in 2 or 3's or as individuals in the scorching sun. They are capable of remaining in the open for hours on end and run for a great distance at a speed of 30-32 miles per hour. The Indian wild ass mate in the months of August, September. The gestation period is about 11 months. Young ones appear in the months of July, August, September. They are called FOALS. They are readily tamed when young. They grow vicious and untameable animals. The present population in India is below 2000.

Rhinoceros Unicornis

Common Name —Great Indian Rhinoceros.

Local Names —Hindi-Gainda, Gargadan; Marathi-Genda.

Size —Average height of male-5 ft. 8 in (170 cms) with a girth of 11 ft. (335 cms) behind the wither.

Distribution —The Great Indian Rhinoceros was formerly extensively distributed throughout the Gangetic plain, but is now restricted to the Dooars and parts of West-Bengal and Assam.

Distinctive Characters —The skin is divided into shields by heavy folds before and behind the shoulder and in front of the thighs. The fold is not continued across the back which is a distinctive character in the Indian species. Another distinctive character is that the hom is smaller in the Indian Rhinoceros.

Habits —Rhinos prefer swamps and grasslands but are also found in wood jungles and ravines of low hills. They lead a solitary life. Rhinos feed on grass. They breed through out the year. Their Gestation period is 16 months. The young ones are born as heavy as 60 kgs and are about 105 cms in length. The female Rhino acquires sexual maturity at the age of 4 years while the male takes 7 years to attain sexual maturity.

The Indian Rhinoceros is a fast diminishing species. They are hunted for sport and their skin is used for making leather products. Their horn is supposed to have medicinal value.

In India, they are found in majority only in the Kaziranga Sanctuary in Sibsagar/Nowgong district in Assam. The Bengal Rhinoceros Preservation Act of 1932 and the Assam Rhinoceros Preservation Act of 1954 were fruitful in protecting the animal and by the end of 1978 the population of the Rhino in the Kaziranga Santuary was estimated to be around 800.

Sus Sulvanius

Common Name —Pygmy Hog.

The Pygmy hog is scarcely 10 inches high (25 cms). They inhabit the forests of the Himalayas in Sikkim, Nepal, Bhutan and Assam. Their habits are similar to that of the wild boar. They live in herds of 5-20 in number. They are nocturnal by nature and rarely seen. Hence nothing much is known about this species.

Cervus Duvauceli

Common Name —Swamp Deer.

Local Names —Hindi—Barasingha, Maha, Central India—Goinjak (male), Gaoni (female), Sal samar, Nerwari.

Size —Average height 54 in (135 cms) at the shoulder, Average weight 370-400 lbs (170-180 kgs), Average length of the horn—30 m. (75 cms) with a girth of 5 in. (13 cms) at mid-beam.

Distinctive Characters —The coat is almost woolly and has colour shades from brown to yellowish brown. Stags are maned and darker in colour. Young ones are spotted. The size and shape of the antlers vary.

Distribution —Swamp deers are found in Terai, Uttar Pradesh, Assam, Sunderbans, Madhya Pradesh.

Habits —They live on marsh land, seldom out of water. They are also found on grassy maidans. They are highly gregarious by nature and less nocturnal. Swamp deers feed till late morning and in the evenings, taking rest during most of the hot day. Their eyesight and hearing are moderate, but their sense of smell is well developed. They are an endangered species and strict protection is necessary to prevent their total extinction. The total number in Central India is less than 200.

Cervus Elephas Manglu

Common Name —Hangul.

Local Names —Kashmir—Hangal, Hanglu (Mate), Minya-mar (Female) Hindi—Barasingha.

Size —Average height—40-50 inches (120-125 cms)

at withers. Average weight-400 lbs (180 kgs). Average length of horn-40 inches (100 cms).

Distinctive
Characters —The coal is light to dark brown fading to a dinghy white on the lips, chin, underparts and buttocks. The white rump patch is divided by broad median stripe extending to the base of the tail. Fawns are spotted.

Tribution —Northern side of the Valley of Kashmir and some adjacent valleys.

Habits —The Hangul is basically a forest animal and leads a solitary life or live in small parties. They roam around in search of food. They give birth to their young ones in late May. They spend their time in summer at heights of 900-12,000 ft. (2750-3680 m). They shed their antlers before they migrate and grow new ones. This is a very regular feature among this species. In winter they descend to lower levels and feed on fresh grass. Another of the endangered species that requires direct human protection.

Cervus Eldi Eldi

Common Name —Manipur Brownanlered Deer.

Local Names —Manipur—Sanga, Burma—Tamin.

Size —Average height 4 ft. (120 cms) at withers, hinds are smaller. Average length of antler—42 inches (107 cm). Average length of head —over 35 inches (90 cms).

Distinctive
Characters —The coat is coarse and sparse. The stags have a dark brown, or black coat. Hinds are light brown or fawn. Young ones are spotted. Antlers are almost erect.

Distribution —Manipur.

Habits —The Manipur Deer avoid hills and heavy forests. They prefer open scrub jungle or flat undulating land between rivers and hill ranges. They live in small herds. Their main food comprises of grass on which they feed in the mornings and evenings. Their speed and their sharp eyesight protect them from their predators. Young ones come in the month of October. Gestation period is about 239-256 days. The present number is only about 100 most of which are restricted to Keibal Lanjao Sanctuary on the shore of Logtak Lake in Manipur. This is the reason that they were given the name Manipur Deer.

Bubalis Bubalis

Common Name —Wild Asiatic Water Buffalo.

Local Names —Hindi—Arna (male), Arni (female), Marathi —Jungli mhais, Kannada—Kadu Kano Kartee.

Size —Average height of the Bull-5 ft. 6 in (170 cms) at shoulder to 6 ft. Average weight—2000 lbs (900 kgs) or more.

Distinctive Characters —The wild buffalo is sleeker, heavier had more robust-looking than their tame counterpart. Their body is slaty black in colour. The newborn are light coloured, almost yellow in colour. Their horns are curved or spread outwards. The horns may be flat, triangular in cross-section, the horns of the cows are less massive and vary less in shape.

Tribution —They are found in the grass jungles of Terai and the plains of Brahmaputra in Assam,

Orissa and Bastar District of Madhya Pradesh.

Habits —Tall grass jungles with food, shelter and water to lie in, form the natural habitat of the wild buffalo. They live in small herds. They feed chiefly on grass. Feeding is done only in the mornings and evenings and sometimes in the night. During the day they lie down in the shade. Wild Buffaloes are bold and most savage by nature. They invade cultivation in search of food. Bulls may attack without warning or provocation. A cow with a new born calf is very dangerous. The chief enemies of this species are man and tiger. Their sense of smell is well developed and they posses good hearing. They have moderate eyesight. Calves appear in tfe months of March, April or May. Wild Buffaloes sometime mate with domestic ones.

Details about only a few species have been mentioned as information about the rest has been hard pressed to find. The above facts have been metioned only as an example as to how one can proceed with collecting data about an endangered species for the purpose of conservation.

Modes of Wildlife Conservation in India

In India, several laws have been passed and sanctuaries and national parks have been established for the protection of the dwindling wildlife. India was probably the first country to enact a Wild Life Protection Act. The Wild Birds and Animals Protection Act was passed in 1887 and repealed in 1912. For game protection in the states, in 1927, the Foreet Act XVI was enacted. Indian Board for Wild Life was estalished in 1952 and this was followed by setting up of Wild life Boards in different States in India. In 1972, a new Wild Life Protection

Act was passed. Under this Act, possession, trapping, shooting of wild animals alive or dead; serving their meat in eating houses: their transport and export are all controlled and watched by special staff (Chief Wildlife Warden and authorized officers). This act prevents hunting of femals and young ones. Under this Act the threatened species are absolulely protected and the rest afforded graded protection according to their state of population size.

In India, nearly 200 sanctuaries and national parks have been established for wildlife management. Santuaries are places where the killing and caputuring of any animal is prohibited except under orders of the authorities concerned. National Parks are set up for preserving flora, fauna, landsscapes and historic objects of an area. Some well known wildlife sanctuaries and national parks of India are.

Kaziranga Wildlife Sanctuary

It was establised in 1926 in the district of Sibsagar, subdivision Jorhat (Assam) on the southern bank of the Brahmputra river. It covers an area of 430 sq. kms of forest grasslands and swamps and supports a fauna of 700 Rhinoceros, besides elephant, wild buffalo, bison, tiger, leopard, sloth bear, sambhar, swamp deer, hog deer, barking deer, wild boar, gibbon and birds like pelican, stork and ring-tailed fishing eagles.

Manas Wildlife Sanctuary and Tiger Reserve

It is located in the district of Kamrup in Assam, covering an area of 540 sq. kms. and is situated at an altitude of 80 metres. The River Manas runs through this sanctuary. It contains the following wild animals: tiger, panther, wild dog, wild boar, rhinoceros, gaur, wild buffalo, sambhar, swamp deer and golden langur.

Jaldapara Wildlife Sanctuary

It is situated in the Jalpaiguri district of West Bengal and covers and area of 65 sq. kms of grassland. Wild life fauna includes animals like rhinoceros fiaur, elephant, tiger, leopard, deer and a variety of birds and reptiles.

Palamau National Park

It is situated in the Daltongunj of district of Bihar and has an area 345 sq. kms. The fauna of this include tiger, Panther, sloth bear, elephant, chital, gaur, nilgai tiger, chinkara, chowsingha and mouse deer. The Flora is thick tropical forests.

Hazaribagh National Park

This national park was established in 1954 in Bihar. It has an area of 184 sq. km of thick tropical forests. The typical fauna of this park includes wild boar, sambhar, nilgai, tiger, leopard. sloth bear, hyena and gaur.

Simlipal National Park

It is situated in the district of Mayuri Bhanj in Orissa and has an area of 2750 sq. kms. It is covered with dense sal forests and has been chosen for Project Tiger. Its typical funna includes tiger, elephant, deer, pea fowl, talking mynas, chital, sambhar, panther, gaur, hyena, and sloth bear.

Chilka Lake

This largest inland lake has an area of 1000 sq. kms. and is about 1000 kms from Bhubaneshwar (Orissa). Its fauna includes waterfowl, duck, cranes, ospreys, golden plover, sandpiper, stone curlews, flamingoes, etc.

Kolameru Bird Sanctuary

This is a small bird sanctuary near Tadepallegudam in Andhra state. It is a breeding place for pelicans and many marine birds visit this place.

Vedanthangal Bird Sanctuary

This small-sized but very old sanctuary extends over a lake

of about 0.30 sq. kms, 85 kms from Madras. Many migratory birds regularly visits thist emporary lake. The regular visitors of this lake are birds like spoon bills, open billed storks, egrets, ibis, cormorant, darter, grey heron, pelicans, snipes and dab chicks.

Guindy National Park

It is situated near Madras city and has mainly chitals and black bucks. A few albinos of black buck are also found here.

Point Calimer Wildlife Sanctuary

This is situated at the southern tip of the Thanjavur district of Tamil Nadu abutting the Palk Strait, Its back water and lagoon is visited by flamingoes and pelicans. It is close to the Vedaranyam forests which has a fauna of numerous blacks chitals and wild boars

Mundanthural Sanctuary

It is situated in the Tirunelveli district of Tamil Nadu and was established in 1964. It covers an area of 520 sq. kms. The Tamaraparani river flows through the evergreen forests of the sanctuary. Panther, tiger, sambnar and chital form the major fauna of this sanctuary.

Periyar Wildlife Sanctuary

Situated in the state of Kerala, this sanctuary covers an area of 777 sq. kms. It was established in 1949 around the artificial lake which arose behind the dam built across the Periyar river in 1900. This sanctuary of great scenic beauty supports a fauna of wild elephants, gaurs, leopards, sloth bear, sambhars, barking deer, wild dogs, wild boars, black Nilgiri langur and water birds like grey hornbills, egrets.

Mudumalai Wildlife Sanctuary

This wildlife sanctuary was established in 1949 in North

Western part of Nilgiris in Tamil Nandu. It is known for its rich forests and diversity of fauna that includes wild elephants, gaur, sambhar, chital, barking deer, mouse deer, four-horned antelope, tiger, panther, bonnet monkey, langur, giant squirrel, flying squirrel, wild dog, jackal, wild cat, sloth bear, porcupine, pangolin, flying-lizard, monitorlizard, rat sanke, python and various birds.

Ranganthitto Bird Sanctuary

It covers in area of 166 sq kms and includes a series of islands in the Cauvery river 15 kms of the Bangalore-Mysore road, near Shrirangapatnam. Its avial fauna includes open bill stock, white ibis, egret, spoon bill, wild duck, pea fowl, night heron etc.

Bandipur Wildlife Sanctuary

This sanctuary was established in 1941 by the then ruler of Mysore (Karnataka State). It is situated 80 kms south of Mysore city enroute to Ootacamund. It has an area of 874 sq. kms and is at an altitude of 1454.4 metres. Its forest is very thick and has plenty of rain fall. Its wildlife fauna includes plenty of gaurs and animals like elephant, leopard, sloth, bear, wild dog, chital, panther, barking deer, porcupine and langur.

Cotigo Wildlife Sanctuary

It is located in South Goa and has an area of 105 sq. kms. It has wet evergreen forest of Elve and Berla trees and supports rich fauna of the following animals, gaur, sambhar, chital. hogdeer, barking deer wild boar, hyaena, panther, leopard, jackal, otter, porcupine, and birds like parakeet lorokeet, woodpecker, kingfisher, bulbul jungle fowl, egret etc.

Bhagwan Mahadev Wildlife Sanctuary

It is located in North Goa and has an area of 240 sq. kms and supports fauna almost similar to Cotigao Wildlife Sanctuary.

Sesan Gir

This famous wildlife sanctuary for the Asiatic lion is situated in Gujarat state, 468 kms from Ahmedabad and 43 kms from Veraval. It has an area of 1295 km sq of semiarid country with patches of thorn scrub and deciduous trees. Its fauna includes Asiatic lion, spotted deer, blue bull (nilgai), fur-horned antelope, chinkara, striped hyena, wild boar, porcupine, langur, python, crocodiles and birds like green pigeon, partidge, rockgrouse etc.

Kanha National Park

This national park was established in 1955 in former Banjar Valley Reserve (Madhya Pradesh). This park has an area of 939.94 sq. kms and includes hilly terrain and streams. It is 175 kms away from Jabalpur and has forests of sal trees. Its typical fauna includes animals like tiger, chital, panther, sambhar, black buck and barasingha.

Tandoba National Park

It is located in Chandrapur district (Maharashtra) and has an area of 116 sq. kms. Its fauna include tiger, sambhar, sloth bear, bison, chital, chinkara, barking deer, blue bull, four-horned deer, langur, peafowl and few crocodiles.

Sariska

This is one of the most beautiful sactuaries situated in the state of Rajasthan near Alwar. It has an area of 800 sq kms. and has dense Dohokara and Solar forests. Its fauna inlude tiger, leopard, spotted deer, jungle cat, four-horned antelope, langur, porcupjne, hedgehog, peafowl etc.

Bharatpur Bird Sanctuary (Koeideo Ghana)

It is located in Bharatpur district in Rajasthan, han an area of 29 sq. kms and harbours all kinds of indigenous birds like nesting water birds, water side birds and migratory birds.

More than 328 varieties of birds including cormorants, spoonbills, white ibis, Indian darters, egrets, painted storks, open billed storks, great black necked storks, etc. Many migratory birds like ducks, geese, siberian cranes, etc., regularly visit this sanctuary. Drier parts of this marshy sanctuary have animals like spotted deer, black buck, sambhar, blue but, wild boar and python.

Sultanpur Lake Bird Sanctuary

This small sized (2 sq kms) bird sanctuary is located in Gurgaon district (Haryana) about 30 kms from Delhi. Its typical avian fauna includes crane, sarus, spot-bill, rudyshell and drake.

Shikari Devi Wildlife Sanctuary

It is located in Mandi district in Himachal Pradesh. It has an area of 213 sq kms and has the following animals: snow leoppard, flying fox, black bear, barking deer, musk deer, chakor, partridge, etc.

Bir Motibagh Sanctuary

It is located near Patiala in Punjab. The fauna includes black buck, blue bull, hog deer, hare, jackal and birds like peafowl, partridge, sparrow, babller, myna, parakeet, pigeon, dove, etc.

Dachigam Wildlife Sanctuary

It was established in 1951 in Kashmir, 26 kms away from Srinagar. It has an area of 89 sq kms and has two levels: upper Dachigam at 3692.3 metres altitudes and Lower Dachigam at 1846.2 metres altitude. It mainly preserves hangul or Kashmir stag, musk deer, leopard, black buck, black bear, brown bear and babboon.

Corbett National Park

It is one of India's most famous wildlife sanctuaries and was constituted in 1935 as the first national park of India. It is situated between Nainital and Gharwal districts in Uttar Pradesh. It has an arca of 525 sq kms and is located within west-south bend of the river Ramganga. It supports a rich and diverse fauna of the following: tiger, panther, sloth bear, hyaena elephant, blue bull, swamp deer, barking deer, Indian antelope, porcupine birds like bulbul wood pecker; barbet, babbler, and bee eater, and reptiles like crocodile, python, etc.

Shivpuri Sanctuary

It is located in Madhya Pradesh and is an asylum for tigers.

Annamalai Sanctuary

This sanctuary was established in 1972 in the southern part of Coimbatore district in Tamil Nadu. It has an area of 958 sq. kms and supports rich fauna of animals like elephant, gaur, sambhar, spotted deer, barking deer, nilgai, Nilgiri langur, Nilgiri tahr, lion-tailed macaque, tiger, panther, sloth bear, porcupine and pangolin.

These are some of the major sanctuaries in India. There are several smaller sanctuaries and national parks that house some of the important species. A sanctuary or a national park may be large or small, but planning should be based on a scientific background. To provide the basic needs for the wild life the following steps may be taken as far as possible for conservational purposes.

1. The forest inside the sanctuary/national park should be left unexploited as far as possible. It would be an ideal environment for the wild life to get the fullest measure of protection, and in such preserved conditions of life they ean multiply freely
2. If total unexploitation is not possible, the cutting operations of the forest should be in small units, well-distributed over the forest. Frequent light cuts are generally better for wild life than infrequent heavy cuts.

3. The forest should not be planted with or allowed to grow only a single species of plant. Further, fruit-bearing shrubs and trees should be encouraged wherever possible and the natural openings of the forests should be preserved.
4. Poaching needs to be ruthlessly prohibited in the sanctuaries in particular and generally in other area outside them, where wildlife still exists. Mobile armed units should be attached to all sanctuaries to patrol regularly the entire sanctuary to stop illegal activites.
5. The water holes and stream sides should be managed and guarded carefully for the benefit of wildlife. Such critical areas may be of small size, but their importance is disproportionately great because they supply important elements in wildlife ecology for larger areas during hot months of the year.
6. Annual burning of the grassland in most of our sanctuaries has become as integral part of management, since not burning these might bring about a lack of suitable fodder during the hotter months. Controlled burning has been found to increase forage and to preserve organic material in the soil. Such burning may be done in blocks with unburned areas in between, so that areas are burnt in rotation. The burning is to be confined to definite blocks by means of five-breaks. Burning at night is favoured unless it is likely to extend the fire beyond the desired limits.
7. Provision for dust bath for animals should also be made be leaving some exposed soil for their dusting. These spots may be prepared in the form of small mounds elevated a few inches from the forest floor to allow for drainage and to dry quickly in wet months.
8. Some artificial salt-licks are also to be provided in the sanctuaries since the wildlife need them at regular intervals for their normal health. Such salt-licks are necessary for animals especially in areas where natural salted clay is scanty.

9. Concentrated grazing by domestic live-stock is to be prevented as wildlife can never compete with domestic stock under ordinary circumstances and domestic animals may be responsible for transmitting diseases like foot and month diseases, rinderpest, surra (sleeping sickness), heamorrhagic sepicaemia, arthrax etc.
10. Cultivation near the sanctuary should be avoided as a rule as it may have an indirect effect on the wildlife. The chemical control of certain pests is often desirable for better yield in agriculture, but these pesticides can present ecological hazards for wildlife.
11. Scientific studies of the sanctuary wildlife particularly of threatened species, by qualified persons, should be encouraged in order to improve the status of the animals even by breeding them in captivity to rehabilitate them in suitable habitats.

Most wildlife could be saved by sufficiently rigid protection of their natural habitat, combined with controlled exploitation of species which are of economic importance.

Some wildlife may be driven away from their natural habitats so that their requirements for not in conflict with human interests. The only possible course for maintaining their viable number is in national parks, sanctuaries, game reserves, where they could thrive unharmed.

In the case of the most endangered species which have reached a point of extinction in their favourable habitat, their survival may be ensured through the maintenance of breeding stock in zoological parks in suitable parts of the country.

Man, An Endangered Species

"Man is part of the vast web of life and cannot escape the natural consequences of his action. . . . The sheer power of the population and technological revolutions may make man himself an endangered species in many parts of the world."

—*Stewart Udall*

The present day presents Man as the most successful species in the world. The ability to think, a sixth sense, is what places the species of Homo sapiens at the top of the ladder. It is this ability that has assisted Man to become the most dominant form of life on earth. In the game of survival, Man has come out in flying colours, this being possible only by his dependence on the environment around him. Initially, the progress of Man was very slow, but with time it has increased to a dizzying pace, this being made possible by the explosion of scientific knowledge and its application to every phase of modern life.

Man's dream for the future is machine oriented civilization. This coupied with his greed is definitely going to have an adverse effect on his natural environment, soul and spirit.

An endangered species is one whose population is decreasing. In Man's case it is the ever increasing population that is making him an endangered, species. There is a general belief—more people technology—which if followed blindly, as is being done, is to flirt with ultimate disaster.

The result of progress and technological advancement of the pollution of water, land, air removal of forests and other natural resources beyond redemption. All these activities tilt the ecological balance on which all life depends.

Each species is a part of a food chain and man is no exemption. Thus preservation of wildlife forms a major decision in the fate of man Man's attitude towards nature only suggests that he is not ready to believe this fact. Unless this attitude changes, all one can decipher is that man is paving roads leading of extinction.

President Johnson of the united States of America in an address to the Congress in 1968 observed that—"Conservation's concern now is not only for Man's enjoyment but for Man's survival." Modern Man, if he is to survive has to improve his relationship with nature. If he succeeds—he must, he will—all wildlife too will have a better chance.

The Future

The future of our wildlife looks very bleak. If wildlife is not preserved then world would be a paradise lost, to put it in better terms, a paradise destroyed. As Man's life becomes more and more urbanized, mechanised and machinised, the habitat for wildlife is being devoured and wild animals are losing their place on our globe. Preservation of habitat is the most important criteria in conservation of wildlife.

With the ever increasing human population, the Idea of conservation can be scarcely brought about. It is a habit for nature to eliminate anything that exceeds a limit and Man, the most powerful and highly developed of living creatures, fails to recognize the needs for his own existence in the future.

Nature and wildlife are of definite economic value and components of an environment of which Man is a part and on which he is directly or indirectly dependent. Man enjoys music, poetry and paintings, all he has to realize is that the flora and fauna are the music, poetry and paintings created by nature. Even more, they form a part of the food chain which is a guarantee for the the continuity of life on earth.

Very few individuals in this world have time for recreation, leisure or intellectual interests. The rest of them think wildlife conservation is a luxury. Each of them is too busy trying his own problems for him to even consider the idea of conservation. Starving illiterates have time to think only of ways to earn their next meal and not about conservation, which for them is non-functional luxury.

Human overpopulation and illiteracy are two of the major problems that bring about the tragic destruction of environment and animal exterminations in the world. For every extinct or endangered species or subspecies of animals, two of the major factors have been the destruction of habitats and excessive cropping by man. Connected with this depletion is the lack of knowledge about the essential needs of a species for survival, its needs for habitat, and its reproductive capacity.

Several organizations are painstakingly to trying bring about a successful turn-about regarding the concept of conservation. They educate and advise on the need for conservation of our depleting natural resources. All these organizations are run by men who are devoted and work their hearts out for saving habitats from destruction and species from extinction throughout the world. They fight against social, economic, political and religious problems in different parts of the world.

It all looks like a hopeful sign that the future of the world's wildlife is of growing concern to more and more human beings. But this is a race against time. Will people and governments be a ware of this before too much is lost ?

The future of word's life is still a big question ?

13

Ecology and the Future of Man

Much of the current controversy about ecology concerns the future. Will man continue to progress in knowledge, health and standards of living, or is he headed toward oblivion in the near future? Is the good life for everyone gradually spreading around the world, and is man coming into better balance with his environment, or is he already on a path of self-destruction.

Other scientists consider these attitudes as overly emotional, fear-producing, and scientifically unjustified. They feel that ecologists cannot support many of their prophesies of doom and gloom, and that they are, in fact, raising unreasonable bogeymen. They point out the great strides in health welfare, education and social advance which have already occured, and they feel that there is a good deal of inaccurate sentimentalism about "the good old days." They note that the good old days had no electricity, no easy transportation, long working hours, very little educational and recreational oppotrunities, and plenty of diphtheria, smallpox, cholera, malaria and tuberculosis. They feel that the alarmist-type ecologists cannot recognize progress when they see it. To this the ecologist replies, "Does progress require polluted rivers and lakes, choking air, birds dying from pesticides, intolerable crowding, ulcers in children, violence on the streets, and heart attacks in young men?" Thus the controversy and arguments can continue almost indefinitely.

In general, ecologists have tended toward pessimistic views of the future for more than 30 years. The first scientific books on the ecologic imbalance of man in the world appeared in the 1930's and 1940's (Sears, 1935; Vogt, 1948; Osborn, 1948). Since these early warnings about the ecologic problem of man, these difficulties have generally increased. Ecologists have found ample evidence for their fears. They feel that current events show that world population growth will not be amenable to significant slowing before the world is severely overcrowded. They have found that man has not learned the lessons of history and has continued to destroy his environment. They believe that the future survival of man will be, at best, a close and hectic race between these destructive forces and an enlightened society which can find the knowledge and necessary for solutions. They now see little evidence that man will really achieve these solutions before it is too late.

Essential Questions for Man's Future

The most difficult and essential problems for man's future lie in broad ecologic and social issues. No two questions are more fundamental or more troublesome to the future of man the ecologists' inquiry.

"What is the carrying capacity of the earth for man? In other words, what are the limits of the life support capabilities of the earth in terms of human population?" are the behaviourists' concern.

"What are the tolerance levels of man to the pressures generated by his own societies? In other words, if man is his own worst enemy, does this lead to his what stimulus, does this lead to his self destruction?"

The carrying capacity, or life support capability, of the earth for man may be dependent upon a great variety of factors.

We can readily identify oxynen water, food space and shelter as vital requirements for human life. There might be many others as well, such as mineral resources, energy availability, or the concentration of toxic metabolites, which could reach states above human tolerance levels. Most of these, however, would operate through the media of air, water or food. Hence, it is worthwhile to examine each of these briefly.

Oxygen Supply and Air Quality

The atmosphere of the earth consists of about 78 per cent nitrogen, 21 per cent oxygen, 0.03 per cent carbon dioxide, and the remainder, other gases. This composition is remarkably stable despite the fact that oxygen is a highly reactive element which readily combines with a great many other elements in chemical processes of oxidation.

The ecologist Lamont C Cole (1970) has pointed out that oxygen has remained stable in the earth's atmosphere because green plants recycle it in molecular form through photosynthetic process. Approximately 70 per cent of the free oxygen produced each year is through to come from planktonic plants in the ocean, and the remaining 30 per cent from terrestrial vegetation of forests and grasslands. Cole emphasized that if we should seriously pollute the sea, as is already happening in coastal areas, and if we should excessively denude the forests and grasslands of the earth, we shall be in danger of upsetting the oxygen balance in the atmosphere. In other words, we must look upon the oceans, forests and grasslands, as the world's oxygen tank, and we must insure their continued healthy ecologic function.

This point of view is contested by W.S. Broecker (1970), a geochemist at Columbia University, Dr. Broecker points out that each square meter of the earth's surface is covered by 60,000 moles of oxygen gas. Plants in the ocean and land produce only about 8 moles per square meter per year, and this is almost entirely consumed by their own respiration and that of animals and bacteria. Thus other organisms use all but

0.0004 per cent of the oxygen produced by plants. Broecker maintains, therefore, that man does not depend upon the oxygen generated by plants, and that there is no current danger at all to his oxygen supply. "In conclusion it can be stated with some confidence that the molecular oxygen supply in the atmosphere and in the broad expanse of open ocean are not threatened by man's activities in the foreseeable future. Molecular oxygen is one resource that is virtually unlimited." Broecker feels that man's existence will be threatened by other types of environmental pollution long before his oxygen supply becomes endangered, and that we need not worry about total oxygen supply.

This is typical of some of the different opinions which are unfolding between physical and biological scientists. Such differences often indicate the need for more research and better knowledge before we can fully accept one position or another.

Another atmospheric problem of long-term concern is the increase in carbon dioxide in the atmosphere. Between 1860 and 1960 the combustion of fuel increased the CO_2 content of the air by 14 per cent of its formerly stable level (Commoner, 1966). Carbon dioxide tends to reflect infrared rays from the earth, and this provides a "greenhouse effect" for the whole earth; that is, it accumulates heat. This will gradually increase the temperature of the earth—an effect which may have long-term implications in climate modification. A rise of just 2 or 3 degrees in annual mean temperature would have a major melting effect on glaciers and ice caps, and this could raise the level of the seas several feet. Such a rise in the sea level would inundate many of the world's major cities. Some experts predict it may require 4,000 years, and others feel it will not happen at all. The latter believe that any occurrence of the greenhouse effect due to increasing CO_2 in the atmosphere will be offset by increasing particulate pollution in the air and increasing capacity of the earth's atmosphere. Thus less light energy will enter, and we could actually have a cooling of the earth's surface. Once again, we see fundamental controversies on major issues largely because we do not really have the complete answer.

The atmospheric problem may also become acute in terms of various toxic elements released into surface, air and water; carbon monoxide, sulphur dioxide, lead from automobile exhaust, hydrocarbons from industrial operations and numerous other products of modern technological societies. For example, 500 million pounds of lead are discharged into the atmosphere of the United States each year from automobile exhausts alone, and significant amounts of this become incorporated into living organisms. A recent study in Sweden showed concentrations of lead in mosses to be as high as 300 to 500 ppm along roadsides. Some roadside plants in the United States may have lead levels up to 3,000 ppm. This lead enters the plants *via* both air and water and can then enter man and animals. A hundred years ago, before the automobile, concentrations in roadside plants were less than 20 ppm. Lead levels over 0.5 ppm have detectable physiological effects in man and animals, and 1.0 ppm levels are often toxic.

In conclusion, there seems to be little or no immediate threat that we will run out of air for breathing, but several warnings have appeared regarding the quality of our air. We apparently have an adequate supply of total oxygen for many centuries to come, though even this is a debatable point. In some localities the situation may be critical at certain times. If stationary inversions of air trap the polluted atmosphere over some of our largest cities such as Tokyo, London, New York or Los Angeles, toxic air pollutants increase to dangerous levels. In August of 1969, a massive still-air inversion occurred in eastern United States, causing serious air pollution in urban areas from New Orleans and St. Louis, north to Chicago and Milwaukee, and east to Washington and Baltimore (Brodine, 1971). At no time was there an absolute shortage of oxygen, but dangerous level of carbon monoxide, sulphur dioxide, hydrocarbons and other air pollutants accumulated in several cities. Twenty million people were affected by varying degrees of eye irritation and respiratory diseases. Fortunately, this great blanket of smog was dispersed by upper air currents in the first week of September, and no large-scale health tragedies occurred. More serious events of this type are likely to occur in urban area unless strong measures against air pollution are undertaken.

Water Supplies

Worldwide water supplies represent a concern for the future of man in terms of both the quality and absolute quantity of fresh water. Many cities have reached or are coming close to their total available fresh water supply and are reaching further and further to obtain new supplies. Los Angeles, for example, is planning tremendous canals and conduits which will bring water from the Pacific northwest over 1,000 miles to the growing megalopolis of southern California.

Two Russian geographers and hydrologists, Kalinin and Bykov, have pointed out (1969) that "The world's water resources form a single entity. The basic hydrologic cycle of the world depends upon solar energy, atmospheric circulation, heat reflectivity and surface retention. The latter is strongly influenced by the biotic community, Kalinin and Bykov point out, on the assumption that the moisture content of the earth remains constant that the total gains and losses from the land and the sea are in balance. Thus, the total precipitation equals the total evaporative loss from land and sea. However, the relative retention time and thus the total amount of water in terrestrial and fresh-water environments can change substantially in relation to the biota and terrestrial storage capacity.

A large proportion of the world's fresh water falls and resides in relatively unpopulated portions of the globe, particularly Siberia and northern Canada. Here, man's activities are limited by other factors, and he cannot readily obtain these vast stores of water Secondly, the matter of water quality is of immedate concern. The principle of water reuse is now becoming accepted, so that people recognize much of their water has been used previously by other municipal systems. Each reuse cycle, however, is increasingly costly in terms of treatment, and each reuse cycle requires disinfectants and other industrial chemicals, combining these problems of distribution and quality, it seems that the water problem of the future is not going to be one of a global water shortage on a planetary basis but one of local crises due to the concentrations of people, industry and water use. Certainly, the immediate concern is finding local water resources of adequate quality to meet the burgeoning needs of urban populations. To the city dweller in New York or Hong Kong whose water tap runs dry, or to the refugee in the Middle East who is dying of thirst, it will be little comfort to realise that 29,000,000 cubic kilometers of fresh water are locked in glaciers and polar ice caps.

World Food Supplies

Nowhere is scientific controversy more evident than on the issue of world food supplies. Some scientists predict famines and mass starvation within the next 10 years. They feel the stage is irrevocably set for millions of people to die of starvation in the 1980's. In fact, we already have famine in many parts of the world, if one defines famine as people dying of lack of food. Even at the height of the Green Revolution in Asia in which new strains of rice and wheat were producing phenomenally larger yields, famine conditions existed in parts of Rajasthan in India; in sections of Burma and in southern Thailand. Starvation is now a grim reality in much of the world and it may well get worse despite the Green Revolution.

Other scientists refute the pesrimism and find evidence for a hopeful outlook in the world food situation. They feel that the Green Revolution which was based on new strains of rice and wheat in Asia and Africa, and which produced a doubling and tripling of yields in one or two years, shows what can be done with modern agriculture. For the first time in many decades, agricultural output increased more rapidly than population growth. India began to approach self-sufficiency in food production, and the Philippiness began to export rice again for the first time since 1900.

A word of caution about the Green Revolution is also necessary. It is quite true that new strains of agricultural crops can sometimes dramatically increase yields, but often these new strains require more careful management, more intensive care, more water, more fertilizer and more insecticides. They may be more vunerable to plant viruses or fungi. Although the plant geneticists try to breed in every possible disease resistance, these new strains do not nave the centuries of natural selection and regional adaptation behind them as do the local strains which the villagers have always used. In West Bengal and Orissa, there were reports that some of the newly introduced high-yield strains of rice were succumbing to a disease that was not affecting the old traditional strains. If this happens to a major extent, the Green Revolution could collapse and a severe blow would be dealt to the villager's faith in the agricultural scientist.

Another factor in the world food picture is one of pure climatic good fortune.

Green Revolution in India was accompanied by the best monsoon seasons in over 10 years. A long series of droughts had broken in southern Asia and adequate rains fell in the right pattern. The possibility exists, therefore, that a good part of the success of the Green Revolution was due to good weather, combined with new strains and more fertilizer. This is a question that will be more adequalely answered in future years to see if increasing agricultural production can be maintained in years of poor rainfall.

In India too much rain came and it fell to early. The wheat crops of northern India were magnificient in April and May just before harvesting and threshing, but heavy premature rains in many areas caused the wheat to rot before it could be threshed. Bihar had serious crop lasses from flooding. Once again, this was a tragic and dramatic demonstration that even the Green Revolution remains dependent on climatic factors outside the control of man.

Throughout all of these arguments on how many people the world can feed, one must ask the realistic question: "Is it necessary that there always be a race between population and food production? Is it desirable to be teetering on the brink of starvation of all times? Is it the best representation of the human condition to be struggling in this manner to feed more and more people at a subsistence lever?"

One also has the impression in reviewing the literature on world food supplies that too much emphasis is placed on future predictions, and not enough on current conditions. We already know that starvation and malnutrition exist throughout the world, even in the most advanced countries. We further-more know that malnutrition of any kind has profound effects upon growth, development and behaviour. Children growing up with malnutrition from early infancy definitely show reduced growth rates and possible retardation of physical and mental development. They furthermore show increased susceptibility to infectious disease. Thus, we return to the issue of the quality of life, and the realization that merely preventing death by starvation does not by any means insure a quality nutritional base for adequate growth and human development. It is apparent that the problem of world food supplies will continue for sometime as a problem of both quantity, and quality and especially one of distribution.

Space on Earth

In 1958 Paul Sears wrote an important essay entitled, "The Inexorable Problem of Space," in which he concluded that "Our future security may depend less upon priority in exploring outer space than upon our wisdom in managing the space in which we live." With a shortage of space crowding in upon us in the 1970's, and 1980's, his words gain validity every year.

The earth's total surface is 196,940,400 square miles of which about 30 per cent or 57,230,000 square miles is land surface. Of this total land surface, approximately 5,000,000 square miles are in Antarctica and are too cold to be easily inhabited except by scientists and explorers 9,000,000 square miles are deserts, and 11,000,000 square miles are in mountoinous regions. Of deserts and imountains, at least one-fourth of their jano area cannot be inhabited, so this eliminates another 5,000,000 square miles of land surface for human habitation. We are left with approximately 47,000,000 square miles of land surface, at the most, for the world's population.

The world now has about 3.5 billion people, and will probably reach 7 billion within the next 35 years. This means we now have about 75 people for every square mile of inhabitable land surface on the face of the earth, and this will approach 150 people per square mile by the year 2000 AD.

As with water, air and food, it is important to distinguish between total or potential amount of any given resource on earth and that amount which is actually utilized or practically available to man, it was pointed out in the sections on water and air that the earth's total amount is apparently adequate for many years to come, but the supply in any given city may not be of sufficient quantity or quality to support the standard of living we would like. So it is with space. Although we theoretically have one square mile of land for levery 75 people (about 9 acres per person), the human population is not evenly distributed. In the great urban concentrations of the world densities often exced 50,000 people per square mile. The amount of living space per person in some cities is less than 20 square feet per person. Theoretically, we could redistribute these crowded people to provide more even spacing around the world. Obviously this cannot be done, and if it were attempted it might wreak hovoç with agricultural production.

The economic and behavioural needs of man are such that he must cluster together, but in so doing, he creates superdensities such as Calcutta, London, New York, Paris, Tokyo-Shanghai and Buenos Aires. The slums of which are definitely deleterious to human health and welfare. Why, we might ask, does man insist on this degree of crowding and then complain about it so bitterly? Why do people continue to pour into the cities of the world when they are already known as centers of population' crime and illness?

Perhaps we need to know more about the real spatial and

social needs of man. What are his requirements for space and privacy, and how do these interact with his needs for social stimulation and group activity? Does man become conditioned to crowding in early childhood and then seek further crowding throughout his life? Does he require the social pressure of a crowd because he feels a compulsion to be in touth with the group and thus be able to predict what the group is going to do next? Have were produced, in our urban centers, self-destructive, systems which so modify the behaviour of their inhabitants that these inhabitants exaggerate their own problems and can no longer achieve rational solutions?

These are all tough questions cannot be answered in simple or easy terms. The answers will require the best intelligence we can muster from many disciplines. Some sorts of answers will certainly be necessary before we can even begin to solve the ecologic and behavioural problems associated with our own use of terrestrial space.

Distributional Inequality

With all of the last four life-support requirements of man—air, water, food and space—we have seen that the major problem of immediate importance lie in distribution more than in total amount. Technologist therefore look upon these problems as soluble by technological means. They feel, they are essentially matters of economics and transportation Behavioural and environmental scientists see them as more basic, however, relating to man's fundamental ecologic behaviour. They feel that certain basic social and ecologic tendencies of man prevent equivable distributions of the earth's resources. Obviously this type of argument can go on forever, and it is perhaps necessary to take a pragmatic approach and look at the current picture for the best glimpse of the future.

Many of the industrial nations of the world, especially the United State, the national northen Europe and Japan, have enjoyed rising prosperity for the last 25 years. In ecologic time, of course, this is but a flash in the pan, but let us assume that it can be sustained for another 25 or 50 years. It is worth examining where the material and energetic support for this prosperity arises.

This distributional inequality is another major symptom of ecologic instability in the world. One would hope for some narrowing in the gap between the haves and the have-nots but unfortunately the opposite trend is still occurring. The wealthy nations of the world are far outstripping the poor nations in their rates of economic gain and human betterment. There are, however, self-corrective mechanisms within this system as, for example, when some segment of the labour market in a wealthy country can no longer compete with its foreign counterpart which does the same work for much less. The world economic enterprise continually strives for the cheapest labour it can find which will do a satisfactory job. Thus, there has been a great expansion of manufacturing in Jupan, Hong Kong, Taiwan and Singapore in industries as diverse as steel, electronics, textiles, toys and optics. The Western factory worker can thereby suffer a decline in his economic security and standard of living while the Asian factory worker experiences for the first time an improving situation. The western industrial worker continues to refuse his own working prospects by striking for higher and higher wages. This will certainly be one mechanism by which wealth is more adequately balanced between nations.

Stil the mass of poverty throughout much of Asia, Africa and Latin America is so great that it is difficult to imagine a painless equilibrium. The differential distribution or wealth remain sharp and represents a type of ecologic situation which cannot last indefinltely without severe consequences and reverberations. To the ecologist it means either a successional stage which is not yet mature, or a system headed for a major breakdown. Economic and ecologic maturity should ideally be in the direction of broader and more equitable distribution of the earth's resources—a narrowing of the gap between the haves and the have-nots, both within and between nations provided the total system does not pass beyond the state of corrective change.

Violence and War

Any ecological system or biotic community can be overloaded to the point of collapse by a variety of factors. Certainly one of the factors for man of most immediate and realistic danger is that of social violence. Either by war or by civil riot, man threatens his own existence.

To maintain a sense of physical security man has built an incredible stockpile of destructive power. The nations of the world spend more than 200 billion dollars per year on armaments—more than on health education, welfare, agriculture, child care and care for the aged combined. Twenty tons of explosives are stored in the world for every man, women and child. Russia and the United States have stockpiled 6,000 nuclear warheads possessing thousands of megatons of explosive force. Each megation has more destructive power than which existed in any of the previous wars of history. The world spends almost 8,000 dollars per year training each soldier and less than 100 dollars per year educating each child.

The question is, of course, how long can man continue to preside over armament like this without using it? And the answer seems to be, not very long. The fact is we do use part of this power in an attempt to achieve social goals. It is being used daily in various parts of the world to maintain territory and dominance ranks. It is justified as a spare and deliberate use in hopes of avoiding the all out effort. How long this can continue is anyone's guess, and the ecologist can claim no more comniscience than the political scientist. The ecologist points out, however, that the ecologic and social pressures which generate international military tension may rise with increased population pressure. It is not likely that man can solve the problems of war and violence until he approaches and understands their basic origins in terms of the behavioural and ecologic relationship between social groups, and the environment.

Adaptive Capabilities of Man

Many scholars feel that one of the great problems of the present and the future is the rapid rate of change which is occurring in society and technology. There is not, however, agreement on the consequences of this rapid change, changing life patterns are too rapid for man to adapt sucessfully. Mankind is so overstimulated by the increase in new sight and sounds, so bombarded by worldwide mass media, and so constantly beset with noval situations and new problems that he will break down psychologically and physically. His responses will be maladaptive and pathologic. He will thus collapse under sensory overload.

It is quite possible that each of these points of view has large elements of truth in it. In the present world one can find many examples of both. Our mental institutions are already filled with those who could not adapt to sensory overload and who experience a personal breakdown of physical and mental health. Yet one can find people are living a sub-human existence-where the quality life is so low that the amount of human pain and misery seen unbearable. Yet man has adapted successfully to these in human conditions.

In many ways, the future is with us now. We can find throughout the world an abundance of high quality life in high quality environments, but we can also find an abundance of low quality life impoverished environments. Many of the questions about the future revolve, therefore, around the question of relative balance and proportions. Which way are we heading and which way will the scales tip?

A major aspect of the future centres around our sense of responsibility to posterity. No representative of posterity is here to speak for that segment of society. How do the present populations of mankind regard their obligations to future generations? Are we willing to be more conservative in our life requirements, and are we, in fact, willing to reduce our rates of personal economic growth to insure a better environment and a greater abundance of resources for future generations? This may be one of the central issues of the next twenty years, and one on which the quality of present and future environments depends.

New Approaches to Environment and Development

Human progress has always depended on our technical ingenuity and a capacity for cooperative action. These qualities have often been used constructively to achieve development and environmental progress: in air and water pollution control, for example, and in increasing the efficiency of material and energy use. Many countries have increased food production and reduced population growth rates. Some technological advances, particularly in medicine, have been widely shared.

But this is not enough. Failures to manage the environment and to sustain development threaten to overwhelm all countriest Environment and development are not separate challenges; they are inexorably linked. Development cannot subsist upon a deteriorating environmental resource base; the environment cannot be protected when growth leaves out of account the costs of environmental destruction. These problems cannot be treated separately by fragemented institutions and policies. They are linked in a complex system of cause and effect.

First, environmental stresses are linked one to another. For example, deforestation, by increasing run-off, accelerates soil erosion and siltation of rivers and lakes. Air pollution and acidification play their part in killing forests and lakes. Such links mean that several different problems must be tackled simultaneously. And success in one area, such as forest protection, can improve chances of sucess in another area, such as soil conservation.

Second, environmental stresses and patterns of economic development are linked one to another. Thus agricultural policies may lie at the root of land, water, and forest degradation. Energy policies are associated with the global greenhouse effect, with acidification, and with deforestation for fuelwood in many developing nations. These stresses all threaten economic development. Thus economics and ecology must be completely integrated in decision-making and law-making processes not just to protect the environment, but also to protect and promote development. Economy is not just about the production of wealth, and ecology is not just about the protection of nature; they are both equally relevant for improving the lot of humankind.

Third, environmental and economic problems are linked to many social and political factors. For example, the rapid population growth that has so profound an impact on the environment and on development in many regions is driven partly by such factors as the status of women in society and other cultural values. Also, environmental stress and uneven development can increase social tensions. It could be argued that the distribution of power and influence within society lies at the heart of most environment and development challenges. Hence new approaches must involve programmes of social development, particularly to improve the position of women in society, to protect vulnerable groups, and to promote local participation in decision making.

Finally, the systemic features operate not merely within but also between nations. National boundaries have become so porous that traditional distiactions between matters of local, national, and international significance have become blurred. Ecosystems do not respect national boundaries. Water pollution moves through shared rives, lakes, and seas. The atmosphere carries air pollution over vast distances. Major accidents—particularly those at nuclear reactors or at plants or warehouses containing toxic materials—can have widespread regional effects.

Many environment economy links also operate globally. For instance, the highly subsidized, incentive--driven agriculture of industrialized market economies generates surpluses that depress prices and erode the viability of the often neglected agriculture of developing countries. Soils and other environmental resources suffer in both systems. Each country may devise national agricultural policies to secure short-term economic and political gains, but no nation alone can devise policies to deal effectively with the financial, economic, and ecological costs of the agricultural and trade policies of other nations.

14

The Ecology of Communities

A biotic community is composed of all populations living within a particular area. The structure of a community consists of its species diversity and the patterns of distribution of those species. Species diversity has two components: richness, the total number of species present, and evenness, their relative abundances. Patterns of distribution include vertical layering, zonation, and a combination of the two.

Richness increases from the poles toward the equator; thus Arctic regions have the fewest species and the tropics the most. The size of the habitat area and the length of time available for speciation both affect richness. Thus, large islands have more species than do smaller islands, and old islands have more species than do younger islands of the same size. Richness can also be affected by interspecific interactions, such as competition, predator-prey relationships, parasitism, and other symbiotic relationships. Any one of these factors can ultimately determine a community's structure. For example, the presence of a predator that feeds upon the more abundant of two competing species tends to ensure that both will remain in the area.

Warmth, moisture, and stability of climate all promote a

community's evenness of species. Evenness is also promoted by biological interactions such as interspecific competition for resources and predation. Most communities have a few dominant species that far exceed, in number of individuals, any one of the others, some of which may be only rarely encountered. Northern latitudes and tropics with alternating wet and dry seasons tend to fit such a pattern. The wet tropics have the largest number of species, each in relatively low abundance.

A community's species are not uniformly distributed throughout it but occur in various patterns. Species are often arranged according to differing physical conditions and other factors within the habitat area, a pattern of distribution called zonation. Zonation is obvious even to a casual observer on a rocky seashore. It is also evident on the sides of some mountains. Each organism tends to become established in a zone corresponding to its niche and to which it is better adapted than other competitors. Because of such factors as competition from similar species and predation, the zone a species occupies is not necessarily the one that provides it with optimal conditions of food, light, and moisture and other resources.

Stratification is a pattern of distribution characterized by vertical layering. Examples include the layers of different species found in soil communities and in mature forests. Each level of soil in a prairie, for example, reveals a different species composition although some species may occur in more than one layer. Such factors as a soil texture, moisture, and nutrient availability determine the soil depth at which a given species is found. Stratification of species in a mature forest is often dramatic. Some species of plants and animals live their entire lives only in the treetops, others only on the forest floor, and others only in intermediate layers.

Among other factors contributing to patterns of species distribution in a community are some of the daily or seasonal activities of animals—such as feeding, migratory, and reproductive behaviour—and social behaviour, such as formation of herds. A community's age is also a significant factor in community structure. For example, young communities are likely to be undergoing the process of ecological succession, different stages of which exhibit different community structures.

ECOLOGICAL SUCCESSION

Aquatic and terrestrial areas from which organisms are removed—either by natural factors, such as volcanism, fire, and flooding, or by human disturbances—are soon repopulated by a variety of organisms. The initial colonization is usually followed by a series of stages, a process known as ecological succession. The entire sequence of stages in a succession is known as a sere, and each stage is a seral stage. One of the more dramatic demonstrations of succession results when a temperate forest has been cleared by deliberate cutting or by a severe forest fire.

Succession in a Woodland Community

Most of us have witnessed the earliest stages of succession of a woodland community. These initial stages occur whenever a cleared field or roadside area goes untouched for 2 or 3 years.

Early stages of succession. Left alone, a denuded woodland area is soon covered by a variety of plants. Some spring up from seeds or roots that survived the cutting of fire; others are blown or brought in from elsewhere. The type of soil and the climate are important factors in determining which plant species will grow and survive. Some of the first plants to spring up are the same herbaceous (nonwoody) weeds found in gardens and neglected vacant lots in the vicinity. For the most part, these are plants that are readily dispersed and quickly grow to maturity.

Given sufficient moisture, some herbaceous species grow to much greater heights than others, providing shade and conserving the moisture of the area. Then, species such as ferns that were unable to survive on bare, unshaded soil get a start. Next, various slow-growing woody plants become established, particularly those capable of withstanding the harsh conditions that prevail in a new community. Ants and flying insects arrive daily, becoming increasingly diverse and numerous as time goes on. These first plants and animals to become established in a cleared area are known as pioneer species.

Within one or two growing seasons, a complex assemblage of organisms may become established in the once denuded area. Birds appear, nest, and begin to feed on seeds and insects. Small mammals may find enough cover to enter the area. Both plants and animals add organic matter to the soil, further conditioning it for other species.

Later stages of succession. As the fastest-growing woody shrubs and seedlings of such trees as sumac and locust spring up, their shade adversely affects some of the pioneer species, often including their own seedlings. Many species that thrived best in the direct sun prove to be poor competitors in shade. Densely shaded areas come to contain rather different plant species than the more open areas.

As the years go by, many of the pioneer plant species seem to disappear entirely. Although their seeds or spores continue to enter the area and may even germinate, conditions are no longer favourable for their growth. But without the pioneer species, some of the species of the later seral stages might not have become established. This point has been one of the widely held, logical inferences drawn from the succession concept since early in this century. However, the view has lately been challenged. While it is clear that the appearance of certain heterotrophic species in early stages of suecession requires the prior presence of certain other species, it is not clear that a high degree of predictability is associated with the later stages of succession. For example, it has been proposed that slow-growing, long-lived tree species are the last to become established and then come to dominate what is known as the climax community because they are slow-growing and long-lived, not because other species had preceded them.

Development of the climax community. The shade provided by large herbaceous and woody plants seems to favour slow-growing tree species. As the years pass, the shorter-lived locust and sumac begin to disappear as coniferous (cone-bearing) trees, such as pines and cedars, take over. Depending on conditions, when the forest reaches the age of 20 to 50 years, it begins to form a canopy some 15 to 20 m above the ground. Except where there are rifts in the canopy, only shade-tolerant species grow well on the forest floor. Then, if the winters are not too severe, the conifers may begin to be replaced by such hardwoods as oak, hickory, beech, and maple. Different insect and bird populations are now found in the area. Stratification is now evident: some bird species nest and seek food only along the ground, some only at intermediate levels in the understory, and others mostly in the treetops. Each of the species now present in usually more specialized in eating only certain types of foods than were the species in earlier stages. Depending on its size and proximity to centers of distribution, the forest now provides food and shelter for larger herbivores, such as deer and moose, and carnivores, such as wolves and bears.

As the forest nears 100 years of age, it begins to achieve a somewhat steady state, described as full maturity. But as much as 200 years or more may be required to reach this stage. At full maturity, succession ceases for an indefinite period, with biomass remaining stable and no new species replacing older ones. The forest is now a climax community.

Characteristics of a temperate forest climax commnnity. In a mature temperate forest, the tallest trees range up to 25 m in height. The interior of the forest has a cathedallike appearance; it may be deeply shaded and cool—even on the hottest days. Only in the spring, before the leaves on the trees of a deciduous forest have grown in, does the forest floor experience a brief period of lush growth. The relatively sparse vegetation of the mature forest floor bears little resemblance to the pioneer community. But here and there, a giant tree may have died or been struck by lightning and come crashing down, perhaps taking some others with it. In the resulting open area, where the sun sends a shaft of light down to the forest floor, a small community of plant species typical of an earlier seral stage appears. Succession begins again in such spots, and in time the gap in the canopy disappears as late-stage species fill it once more.

The balance between production and Respiration. In early seral stages, the production of biomass in a community for exceeds the amount needed for maintenance, as represented by the amount of respiration it conducts. Put another way, the community's biomass gradually increases during succession. Once maturity is reached, production of new biomass tends to equal the amount of material used in maintenance and undergoing decomposition.

Lake succession. Succession also occurs in lakes and ponds. The virgin, or uncut, forests of temperate zones long ago reached maturity, following the last retreat of the glaciers, while some of the lakes continue a slow progress toward an equilibrium stage. At first a lake's bottom is rocky and without sediments. The nutrient content of the water is low and support only sparse growth of plants and phytoplankton, a condition termed oligotrophic (from the Greek oligo, "few" or "little" and trophos, "nourishment"). As each year passes, a lake may produce a little more vegetation than is consumed or decomposed. This additional material dies and settles to the bottom along with soil particles carried into the lake by runoffs of rain and melting snow. Each year, such a lake becomes silghtly shallower and slightly smaller in area as benthic, or bottom, sediments accumulate (Fig. 14.1). If the process continues and the lake has a drainage outlet, it becomes a marsh; without an outlet it becomes swamplike. In either case, it is finally overgrowed by the tree species characteristic of the surroudding forest. The soil may remain boglike, however.

Humans have lately played an unintentional role in accelerating the succession of many lake communities and in initiating succession for lakes that have shown no signs of change over a very long period of time. When nutrient-laden sewage effluent and runoff from heavily fertilized farms and densely populated feedlots pour into lake in great quantity, the lake, thus fertilized, becomes an enormous culture vessel for photosynthetic organisms. Whatever the source of the nutrients, this process is known as eutrophication (from the Greek eutrophos, "well nourished") and results in a prodigious growth of algae and aquatic plants. When decomposition of this dense growth occurs, aerobic bacteria deplete the water's oxygen, producing massive fish die-offs. When production greatly outstrips consumption, the lake rapidly becomes a marsh, swamp, or bog. Thus, a process that takes many centuries, or in many lakes appears not to occur at all, is sometimes driven to completion within a few decades. This process has been avoided in certain managed lakes by controlling the rate of fertilization and the harvesting of fish.

MAJOR AQUATIC COMMUNITY TYPES

The following somewhat detailed accounts of nutritional and other aspects of a seashore community and three types of open-ocean communities illustrate the general principles that apply to all community types. The more condensed descriptions of freshwater aquatic communities and of the major terrestrial community, or biome types should be read in light of those principles. One of the principles illustrated is zonation. Another is the dependence of the producers of most communities on the nutrient supply released by decay of dead organisms and on the presence of sunlight, an exception being the chemolithotrophic producers living near volcanic vents on the sea floor. Finally, a dramatic example of the effect of seasonal cycles on productivity is seen in the description of aquatic life of the Antarctic (Fig. 14.1).

Seashore Communities

The community zonation often evident on a rocky seashore has already been noted. Kelp and other seaweeds can be observed in waters where rocks furnish attachment surfaces. In the littoral, or intertidal, zone of a rocky shore—that is, between the

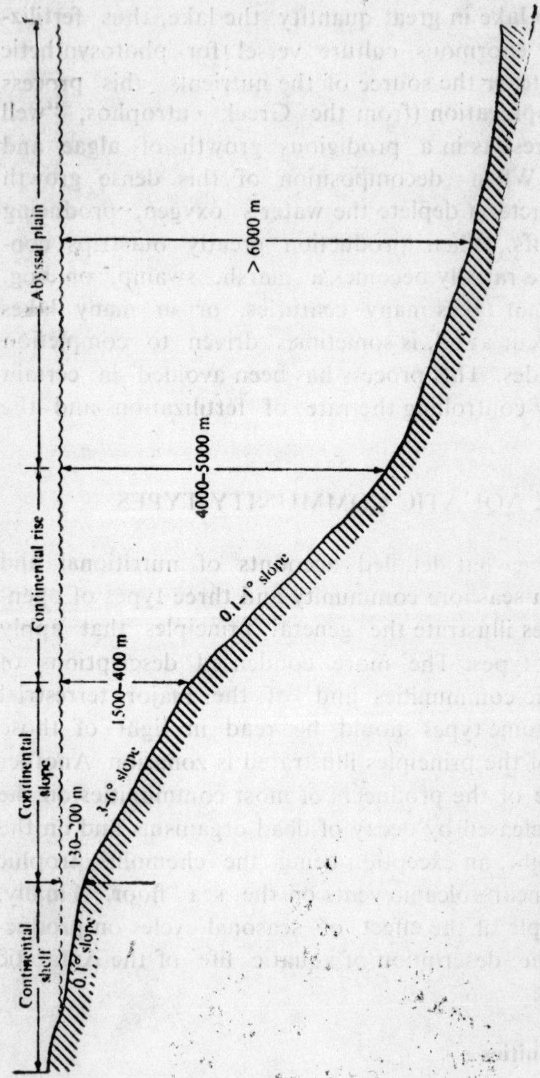

Degrees of slope and distances in meters shown on this drawing are not to scale.

Fig. 14.1: The marine realms diagrammatically represented (not drawn to scale).

high and low tide marks—as well as just above this line, in the splash zone, a series of distinct subdivisions can be recognized. Different species of organisms, each highly specialized for surviving under particular conditions, are found in each of the smaller zones; some of these species. Such organisms are adapted to withstand alternating exposure to water and air or the pounding or sweep of waves. Below the low tide mark are plants that can thrive only without ever being exposed to air. Only organisms that can withstand brief, intermittent exposure to air are found just above the low tide mark.

In addition to the profusion of plants that often meets the eye at the sea's edge is a more important group of producers, ones we cannot see without a microscope; these are the photosynthetic protists, or phytoplankton. Most phytoplankton are diatoms, the one-celled algae enclosed in minute two-part silica shells. Another abundant group, the dinoflagellates, may at times outnumber the diatoms.

The edge of the sea is much more productive than the open seas. The seashore community is the first to receive the rich contribution of minerals and organic matter contained in runoff from the land. Moreover, in relatively shallow water, none of the minerals released by the decomposers are lost to the water below the euphotic zone, the upper, sunlit layer of water in which photosynthesis can occur. Instead, these minerals are immediately availabls once again to the local producers, the phytoplankton and larger algae, ensuring that productivity is as high as light and temperature will allow. For these reasons, the warm waters of tidal flats, and to a lesser extent the shalower regions of the continental shelf, are home to a lesser extent the shallower regions of the continental shelf, are home to a great abundance of life. Such high productivity also characterizes coral reefs. In the open ocean, however, much nutrient material settles into the depths and is lost to the surface waters, greatly reducing their productivity.

Among marine animals, coelenterates, molluscs, annelids, arthropods, and echinoderms predominate at the seashore.

Most of them feed on plankton, algae, or microcrustaceans, those invisible to the unaided eye. Snails rasp algae from hard substrata, while bivalve mollusecs consume bacteria, plankton, and other nutrient materials suspended in the water. Many crab eat algae and other seaweeds, as do many fishes. The tinycrustaceans known as copepods and a variety of animal larvae abound, feeding on bacteria, plankton, and detritus.

As an adaptation to the many types of food available, the animals of seashore communities exhibit a great variety of feeding methods, each of which is an adaptation to a specific ecological niche. One general class of nutrients consists of the fine particles of organic material, living and dead, suspended in the shallow waters at the sea's edge. This constitutes a major source of nutrients and a large variety of animals are adapted to feeding on these particles.

Filter feeders. Fine materials are most efficiently removed from suspension by some method of filter feeding. What is removed from the water and eaten depends on the method employed. In many filter feeders a current of water produced by the beating of cilia directs the food-laden water toward the mouth. As the current bearing the food particles approaches the mouth, a sorting mechanism usually routes particles of the right size and density to the mouth and sends the others away. Feeding currents are used by such diverse organisms as ciliate protozoans, sponges, corals, clams, and the minute, ciliated metazoans called rotifers (Fig. 14.2). In one rotifer, sorting is based on whether the particles are small enough to drop into a ciliated groove (Fig. 14.2). Those that are small enough are passed to the mouth, and the larger ones are rejected. Various filter feeders' are specialized for eating particles of different sizes.

In place of cilia, the copepods shrimps, barnacles, and other small crustaceans that filter feed have fine bristles on their appendages that produce feeding currents and trap food. In bivalves, such as oysters and claims (Fig. 14.3), the work of the cilia is aided by a sheet of mucus produced by the gills. This

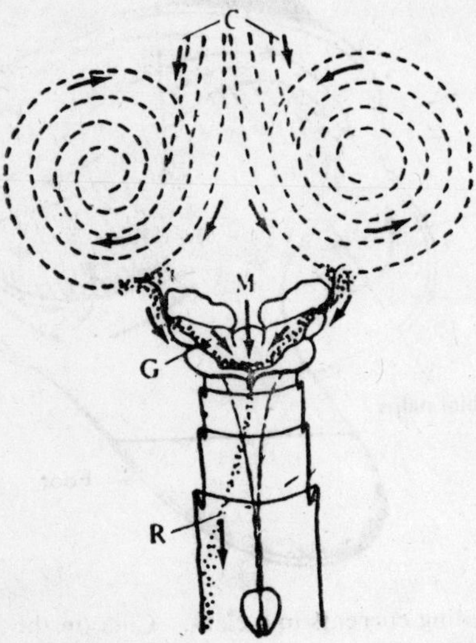

Fig. 14.2: Feeding currents of a rotifer (C=current of food toward mouth; G=groove; M=mouth; R=rejected particles).

mucus serves to trap the finest food particles, including bacteria. Both the mucus and its entrapped food particles are digested, much as humans digest the proteins in the saliva we swallow with our food. Heavier sand grains and other such particles are sorted out by weight before reaching the mouth; they drop to the bottom of the mantle cavity, from which they are removed. Many types of clams extend a tubelike siphon up into the food-laden water above them while lying buried in the sand (Fig. 14.4). Some tub-dwelling annelids; such as the parchment worm (Fig. 14.5), trap fine particles by means of a bolloon of mucus, which they then ingest.

Deposit feeder. Although filter feeders remove much of the suspended matter from the water before it settle to the bottom, bottom dwellers, such as sand shrimp, feed on the rich deposits of detritus and microorganisms that reach the bottom; the process is termed deposit feeding. The material that escapes the bottom dwellers and becomes mixed with the sand and mud

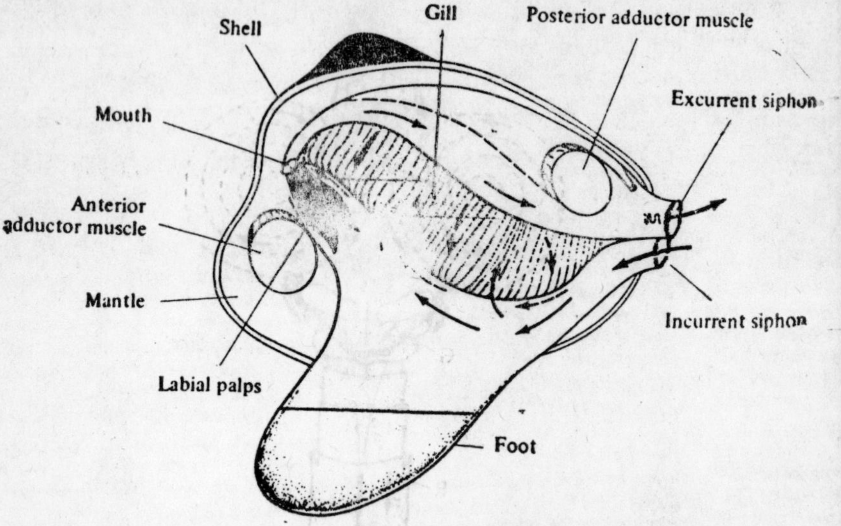

Fig. 14.3: Feeding currents in a clam. Cilia on the gills and palps carry the food particles toward the mouth (indicated by solid arrows); rejccted particles are removed (indicated by dashed arrows).

Fig. 14.4: Some common types of marine clams engaged in filter feeding. Feeding is conducted while tide is in and discontinued when tide is out.

meets another community, the burrowing animals. Many of these eat the entire mixture of food particles and mud or sand and digest the organic matter it contains.*

Among the larger burrowers of the seashore are many kinds of worms, such as the deposit-feeding lunworm. A temperate zone population of lugworms can process some 2,000 tons of

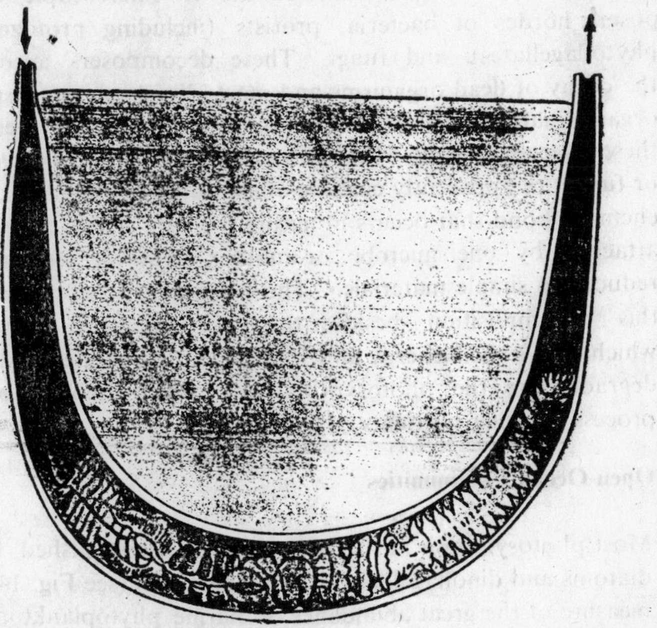

Fig. 14.5: Feeding current of a parchment worm. The worm attaches to the inside of its tube by means of suckerlike appendages and then creates a current by undulating its fans, which are greatly modified posterior segments.

beach sand and mud per aere per year, thereby altering its character and its suitability as a habitat for other organisms. Like earthworms lugworms eat the soil, digest the nutrients it

*Not all burrowers are deposit feeders. Many take advantage of the protection afforded by burrowing but rise at high tide to eat as filter feeders at the sand's surface. Others, including many clams, filter feed through siphons extended up through the sand. Many predators burrow through the sand and mud, seizing eating worms, crustaceans, and other burrowers encountered in their foraging.

contains, and eliminate the residue. In sandy areas, they usually dig a U-shaped burrow with two openings; the sand and detritus that fall into the entrance are eaten, and the undigested residues are discharged *via* the burrow's exit.

The special roles of the decomposers. Most of the nutritional specialists of the seashore are its microscopic decomposers: hordes of bacteria, protists (including protozoa and phytoflagellates), and fungi. These decomposers accomplish the decay of dead organisms and most dissolved or suspended organic matter. The mud and sand of the seashore teem with these microorganisms. One or more types of protists, bacteria, or fungi are apparently capable of splitting every kind of organic chemical bond that occurs in nature. Every type of bond is attacked by one microbe or another until all molecules are reduced to simple materials. The immndiate beneficiaries of all this decomposition—besides the microorganisms themselves, which derive their energy and substance from the materials they degrade—are the photosynthesizers, which depend on this process for their nutrients.

Open-Ocean Communities

Most photosynthesis in the open sea is accomplished by the diatoms and dinoflagellates of phytoplankton (see Fig. 14.1). A measure of the great abundance of marine phytoplankton, each individual of which is no larger than the smallest dust mote that can be seen in a bright ray of sunshine, is provided by the bioluminescent species of dinoflagellates. These cells emit a brilliant white flash of light that lasts for less than a second. Sparkling breakers crashing on a reef or shore on an overcast night in late summer provide a breathtaking sight. Dinoflaggellates left on the beach by a wave light up again when new waves strike them. People walking along the wet sand leave glowing footprints behind. The bodies of swimmers glow brightly, often in various hues. Although dinoflagellates are especially numerous in nutrient-rich coastal waters, a demonstration of their abundance in the open ocean can be witnessed from the deck of a ship under way on a summer night. From either side of the bow there tumbles a sparkling, silverg effervescence, and the ship's wake persists as a broad, silvery highway, easily visible at a distance. The beautiful sight may be seen night after night across thousands of miles of open sea.

Copepods as harvesters. Feeding on the phytoplankton are the zooplankton: colourless species of dinoflagellates, many other protozoans, a host of invertebrate larvae, and large numbers of newly hatched fishes. However, the most abundant of the zooplankton harvesters of phytoplankton are the tiny, filter-feeding copepods, of which Calanus is one of the most common (Fig. 14.6).

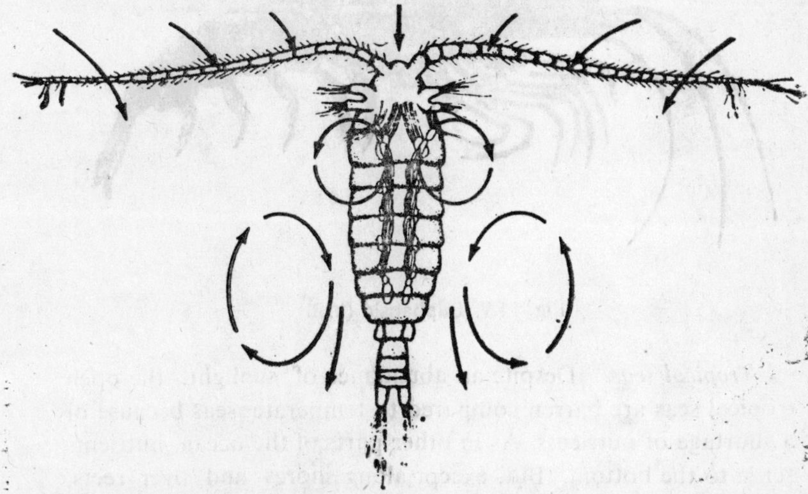

Fig. 14.6: The feeding currents of the most common of the copepods, Calanus. The currents are created as the animal swimgs.

Copepods typically harvest phytoplankton by setting up a current with three pairs of head appendages. These vibrate very rapidly—up to 40 times per second—in a rowing motion that propels the copepod (Greek for "oar foot") smoothly through the water. The inner portion of the current created on either side of the body is directed across a mesh of bristles that trap plankton. From time to time, the collected microorganisms are scraped off and then passed to the mouth by other appendages.

Filter feeding by zooplankton in a sense performs the task of harvesting and processing the phytoplankton for the larger animals, such as fishes, that harvest copepods. For large animals to live by straining phytoplankton from seawater would usually involve the expenditure of more energy than the could obtain from their food. In nutrient-rich, highly productive regions of the sea, the role of the copepod is assumed by larger primary consumers, such as the shrimplinke crustacean euphausid (Fig. 14.7) of Antarctic waters and the small fish anchoveta in areas of upwelling.

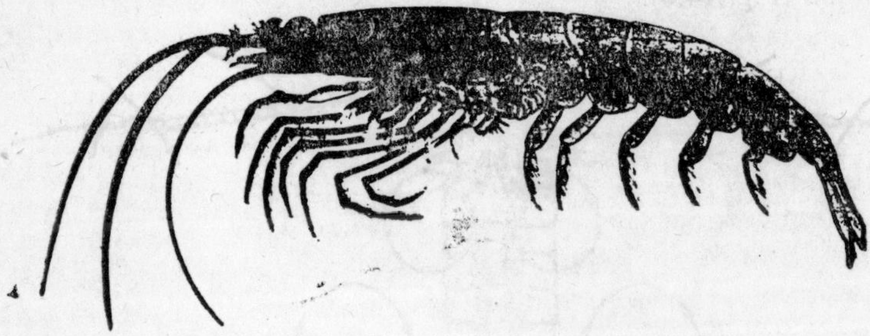

Fig. 14.7. Euphausid (krill)

Tropical seas. Despite an abundance of sunlight, the open tropical seas are barren compared to temperate seas because of a shortage of nutrients. As in other parts of the ocean, nutrients settle to the bottom. But, except along shores and over reefs, only a limited return of these nutrients occurs. Seasonal gales that might bring up mineral-laden water from below are generally absent in the tropics. As in many other regions, a discontinuity layer, or thermocline; usually prevails—a sharp drop in temperature occurring in a zone between a layer of warm water that overlies colder, denser water below. Because layers of two different densities seldom mix or exchange solutes, the upper layer is deprived of minerals. The growth of phytoplankton is therefore limited by whatever nutrients are present in the surface layers. The nutrient-rich deeper waters lie below the euphotic zone, in which sunlight effectively promotes photosynthesis. The result is a relatively scanty crop for the deep blue open seas of the tropics. Nutrient-poor waters are typically clear and, like the sky, a deep blue in colour; shallower, productive waters near shore, over a reef, and in an area of upwelling are cloudy and, due to the presence of phytoplankton, a distinct green.

Life at the Bottom of the Sea

Because living matter and its products are generally denser than water, planktonic organisms tend to sink. Both phytoplankton and zooplankton have evolved a variety of hairlike spines and bristles or other buoyand devices. There add much to the organism's surface area but little to its mass, thereby decreasing its overall density and tendency to sink. Even the gentlest currents thus suffice to keep these tiny organisms near the surface of the ocean, in obedience to the same principle by which a dandelion plume is easily wafted about by a gentle breeze. But just as the dandelion plume falls eventually to the ground, all plankton that is not eaten finally falls to the ocean bottom, where it slowly decays. The detritus derived from the death and decay of nonplanktonic organisms also falls to the bottom.

The bacteria that accomplish most of the decay of this material in the dark, icy depths release a wealth of nutrient minerals into the deeper layers. All across the open oceans, covering 70 per cent of the earth, the surface layers continually lose nutrient materials to the lower levels of the sea. Only in areas of upwelling and in regions subject to violent seasonal storms are some of these valuable minerals regularly returned to the surface, where most of them are once more incorporated into phytoplankton.

Animal life in the depths. Despite tremendous pressure, near-freezing temperatures (2 to 4°C), and a complete lack of sunlight, the depths of the sea support animal life. The ultimate source of food for nearly all deep-sea animals is the "rain" of dying, dead, and decayed plankton and detritus that drops down to them. As in other communities, some of the animals filter feed, some eat deposits, and others prey on the filter and deposit feeders. But so little particulate matter reaches the bottom that only sparse populations can be supported by it. From concentrations as high as 10,000 kg/km^2 of animal life in shallow water, the density of animal life, drops to an average of 520 gm/km^2 far out on the continental shelf. Beyond the shelf, the concentration of animal life averages only 2.5 mg/km^2, although it can be remarkably abundant on the continental slope in regions where the surface waters are highly productive. However, the diversity of bottom-dwelling, or benthic, animals increases with depth. This phenomenon is related to the far greater stability of conditions in deeper waters and the great area of the abyssal plain.

Predominant benthic animals include echinoderms such as starfish, sea urchins, and sea cumcumbers. One entire class of sponges lives only in deep water. A variety of deep-sea nekton, such as fish and squids, is also to be found. In 1960, the Swiss oceanographer Jacques Picard was surprised to see a fish 30 cm (1 foot) long swim by the window of his diving bell Trieste as it settled in the deepest part of the sea, the bottom of the Mariana Trench, 11 km (7 miles) below the surface.

Hydrothermal vent communities. What some have called the oceanographic discovery of the century was made in 1977 by a Woods Hole Oceanographic Institute (WHOI) expedition to the Galapagos Rift, in the Pacific Ocean. The rift is 2,500 to 2,700 m (over 1.5 miles) below the surface, 380 km (236 miles) northwest of the Galapagos Islands, and 1,000 km (620 miles) west of Ecuador. The rift is part of an extensive global midoceanic ridge system that has developed where the tectonic plates of the earth's crust are moving apart. In such places, a flow of lava occasionally emerges and hydrothermal vents spew out hot water rich in H_2S and other minerals.

The WHOI expedition of 1977 and a 1979 expedition using very sophisticated equipment studied the Galapagos Rift hydrothermal vent ecosystem in detail, measuring the temperature and the mineral content of the water and photographing and collecting specimens of more than a dozen species of organisms found around the vents. Whereas water temperature in the sea's depths is usually 2.0°C, it was typically 10 to 15°C at hydrothermal vents along the rift. Some chimney like volcanic vents where the water reached 300°C ware also discovered. One unexpected finding was that the life of a vent community is based, not on the "rain" of material generated by the producers in the euphotic zone above, but on a rich community of chemolithotrophic bacteria that derive all the energy they need from oxidation of inorganic compounds such as the hydrogen sulphide that emerges from the earth's interior. This discovery was truly revolutionary. The bottom of the sea has long been classified as an incomplete ecosystem because it lacks photosynthetic producers. This view must now be modified for the regions around hydrothermal vents; although no photosynthesis occurs there, a thriving population of chemolithotrophic producers does.

Bacterial counts for water samples taken 1 m above vents were 10^5 to 10^6 cells/ml. Around each active vent, the bottom was covered with hundreds of clams, mussels, enteropneusts (a type of hemichordate), and many large, tube-dwelling worms. These worms, some 1.5 m long, live only in the warm water, but most of the animals species live in the cold (2.0°C) water nearby. Animals thus far encountered are unusual and include a new phylum of worms. One species of crab can survive only under pressure at least 125 times atmospheric pressure.

Antarctic Seas

The ward Antarctic usually conjures up throughts of sunless winters in which 40° below zero—on either the Celsuis or the Fahrenheit scale—would be a warm day. Yet Antractic seas support much life. Great mats of cyanophytes, or blue-green bacteria, are sometimes seen beneath the ice. Fish are plentiful, and several kinds of birds, seals, and whales thrives on them. The waters are usually quite rich in nutrients, the key to their fertility. Unlike Peruvian waters, however, there is a sharp seasonal fluctuation in productivity due principally to a lack of sunlight during much of the year.

The annual cycle. In the Antarctic spring, melting ice and snow create a slow surface current of dilute seawater moving toward the equator. This creates an equally slow countercurrent of denser (because it is saltier) mineral-laden water just below it moving toward she South Pole and rising to the surface to replace the water that moves away. The bright spring sun shining on the mineral-rich water produces a bountiful harvest of diatoms, whose numbers double daily for several weeks (Fig. 14.8). By early summer, the shrimplike euphausids, which feed on diatoms, increase to the point at which the sea is a thick soup of krill, as the Norwegian whalers call these small crustaceans, which they encounter in northern waters. But then the phytoplankton begin to decline. both because they have been eaten in large numbers by the euphausids and because their nutrients—nitrates and phosphates especially—have begun to be exhausted. A succession of different types of zooplankton and fishes follows, with larger animals feeding on smaller ones and the young fishes increasing in size daily.

As fall makes its early appearance and the slanting rays of the southern sun no longer warm the waters, the surface layer cools and mixes with the lower layers. Fall gales further mix the deeper mineral-laden water with the surface layer, and diatom growth resumes, but only briefly and never reaching its springtime peak (Fig. 15.8). Soon, falling temperatures and the darkness of the early winter halt photosynthesis. The krill no longer congregates at the surface. Bnt the surface layer is already fertilized and ready for the spring bloom when the sun rises high in the sky again the following year.

Antarctic food pyramids. In Antarctic seas, as elsewhere, birds are often the dominant animal atop food pyramids. But two other, strikingly different food pyramids exist in the Antarctic. In one, the killer whale, which feeds on porpoises and seals, occupies the top position (Fig. 14.9). Killer whales are voracious eaters and have been known to swallow 32 full-grown

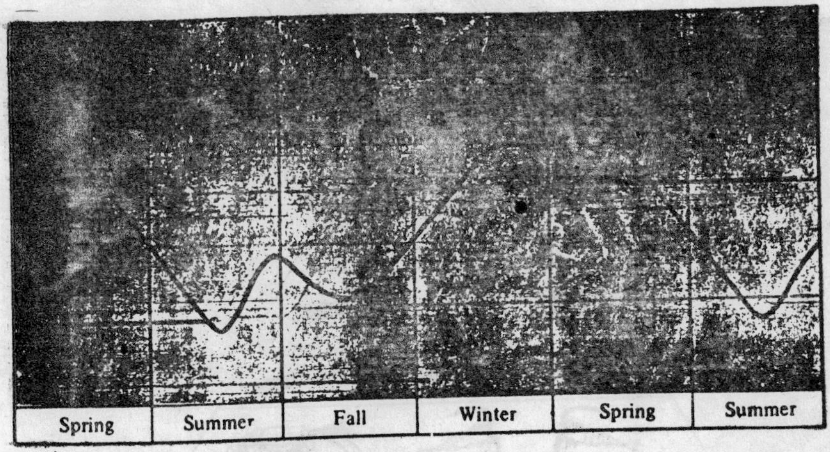

Fig. 14.8: The annual Antarctic sea cycle. Violent winter storms and the movement of melting ice and snow away from the poles in the spring bring up mineral-laden water from the depths. The fertile seas produce an enormous diatom harvest by late spring, ending when nutrients are depleted. When nutrients increase again in late summer, a second but less extensive diatom increase occurs. This inerease stops as the early Antarctic winter begins.

seals in one meal. This is expensive food in that, for every 10 kg a killer whale grows eating such food, up to 100,000 kg of phytoplankton are needed at the bottom of the pyramid. This, of course, puts a low practical limit on the number of killer whales that can be supported by eating seals alone. The baleen, or whalebone, whales, which include the humpback and the blue whale, the largest animal that has ever lived on earth, have evolved a different approach to nutrition. Full-grown blue whales exceed 30 m (98 feet) in length and 150 tons in weight. What food could support the life of such a creature? The answer is something of a surprise.

Baleen whales are named for the fringes of baleen, or whalebone (not true bone), that hang down in mustachelike sheets inside their mouths (Fig. 14.9). Some species of these whales, including the blue whale, swim through dense schools of krill,

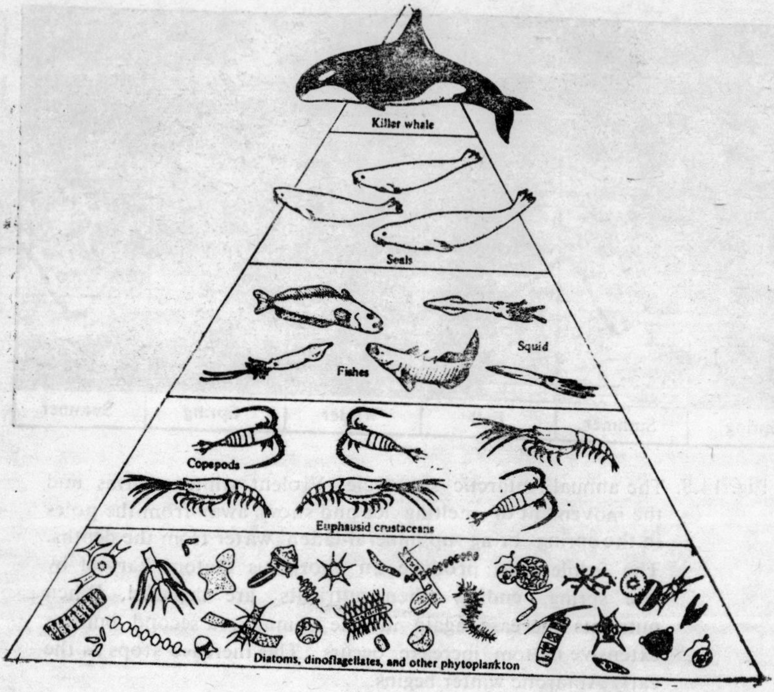

Fig. 14.9: The killer whale food pyramid. For a killer whale to gain 10 kg., up to 100,000 kg. of phyt oplankton must be grown and harvested. At each level of consumption, up to 90 per cent of the mass and energy of the food organism is used to maintain the consumer.

straining huge amounts of water through the baleen mesh-work. A large whale's expandable mouth holds 60 m^3, or 60 tons, of water, which it filters of krill in a few seconds. The filtering process allows the phytoplankton to escape with the seawater while the krill is retained to be swallowed at leisure. Tons of krill can be eaten by a whale in a short time. By eating primary consumers directly, the blue whale and several other baleen whales have cut out the intermediate levels of the food pyramid, a most efficient approach to nutrition. Thus, 100 million kg of phytoplankton production supports up to 1,000 kg of blue whale growth, compared to only 10 kg of killer whale growth (Fig. 14.10). This efficiency has provided enough food to support the growth of individual blue whales to enormous size and the growth of their population to hundreds of thousands. However, in recent years, due to severe over-hunting, populations of the blue whale and several other whale species have dropped so low that reproduction has become affected and some species are not expected to survive.

Freshwater Communities

Compared to marine and terrestrial communities, freshwater

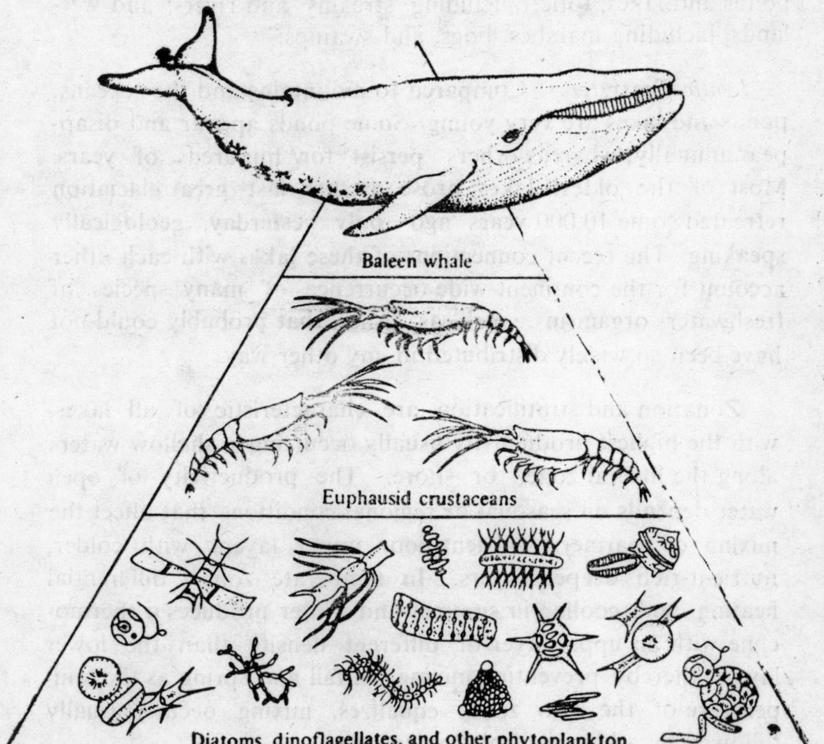

Fig. 14.10: The baleen whale food pyramid. The hump-backed whale, a type of baleen whale, eats krill directly. Two higher orders of consumers found in the killer whale pyramid are thereby climinated, resulting in great nutritional efficiency.

ecosystems occupy a rather small area of the earth's surface. Their impostance is quite out of proportion to their size, however. Because water is so resistant to temperature change, the presence of large bodies of water prevents drastic daily fluctuations in temperature on nearby land. Moderation of seasonal temperatures occurs for the same reason. Because rain and melting snow that soak into the ground reappear at lower altitudes in spring-fed streams, marshes, swamps, pounds and lakes, such bodies are an integral component of the hydrologic cycle. Freshwater ecosystems are classified as lentic, including ponds and lakes; Iotic, including streams and rivers; and wetlands, including marshes, bogs, and swamps.

Lentic Ecosystems. Compared to mountains and the oceans, ponds and lakes are very young. Some ponds appear and disappear annually, whereas others persist for hundreds of years. Most of the oldest lakes arose as the last great glaciation retreated some 10,000 years ago—only yesterday, geologically speaking. The recent connections of these lakes with each other account for the continent-wide occurrence of many species of freshwater organisms, such as fishes, that probably could not have been so widely distributed in any other way.

Zonation and stratification are characteristic of all lakes, with the highest productivity usually occurring in shallow waters along the littoral zone, or shore. The productivity of open water depends on seasonal or regional conditions that affect the mixing of warmer, nutrient-poor upper layers with colder, nutrient-rich deeper layers. In temperate zones, differential heating and cooling in summer and winter produces a thermocline with an upper layer of different density than the lower layers thereby preventing mixing. In fall and spring as the temperature of the two zones equalizes, mixing occurs, usually followed by a bloom of phytoplankton.

Lotic Ecosystems. Running water, such as that found in streams and rivers, has several special characteristics. First, the oxygen content of streams and rivers is usually high and the proportion of their nutrients obtained from adjacent lands is great. Depending its velocity of flow, moving water may severely limit the kinds of organisms that can survive in it. Small, rapidly flowing headwaters, or upper tributaries of a river, are often shaded from sunlight and therefore poor in productivity and species diversity. The consumers of headwater communities depend largely on detritus, leaves, insects, and other organic materials that fall into them or enter them with runoff. In contrast, the wider midsections of large rivers are characterized by a high productivity of algae and a predominance of filter feeders and other primary consumers. Productivity and species diversity both reach their peak in these midsections. Long rivers become increasingly muddy, which reduces the penetration of sunlight, and hence the productivity and species diversity that depend on it. The rate of productivity in muddy waters is lower than the rate of consumption. Rivers slow as the approach their mouths, resulting in the settling out of many of the heavier particles of silt and sand. The great volume of sediment that accumulates near the mouth of a large river forms a delta. The "toe" of the boot-shaped state of Louisiana was formed in this way.

A great variety of conditions can be found along the length of a river. The shallows of a rapid river can be as quiet as any pond. Where current velocity slows in deep pools, sufficient sand may settle out to provide a habitat for various burrowing animals. During flooding, a river may directly connect with marshes or other wetlands, with which it exchanges nutrients and various species of organisms.

Freshwater Wetlands. The two major types of freshwater wetlands are marshes and swamps. Swamps are populated by trees, whereas marshes are not. Wetland sediments of marshes and swamps form two principal layers, the uppermost of which is a relatively shallow zone in which production occurs and oxygen is sufficiently abundant to support aerobic decomposers. A much deeper anaerobic zone of muck lies below the upper layer, accommodating several types of anaerobic bacteria that reduce organic matter to such gaseous forms as carbon dioxide (CO_2), molecular nitrogen (N_2), hydrogen sulphide (H_2S), and methane (CH_4). These gases rise to the surface and are released into the atmosphere, thereby completing the cycle begum when they were incorporated into organic matter. Hydrogen sulphide is responsible for the "rotten egg" odour of some marshes, and methane, when ignited by spontaneous combustion, accounts for the pale blue flames, called willo'-the-wisp, sometimes seen in marshes and swamps at night.

Although coastal and freshwater wetlands occupy only about 2 per cent of the earth's surface, their importance in recycling the world's sulphur, nitrogen, phosphorus, and carbon is quite out of proportion to the area they occupy. Wetlands are also breeding grounds for many animal species and feeding grounds for billions of migratory waterfowl. Only recently have public attitudes towards wetlands begun to change from regarding them as wasted areas that should be drained and "improved" or commercial purposes to the realization that they are valuable resources worth more to us in their natural state than as agricultural or commercial developments.